DISPERSAL CENTRES OF SPHINGIDAE (LEPIDOPTERA) IN THE NEOTROPICAL REGION

BIOGEOGRAPHICA

DR. W. JUNK B.V., PUBLISHERS, THE HAGUE-BOSTON 1978

DISPERSAL CENTRES OF SPHINGIDAE (LEPIDOPTERA) IN THE NEOTROPICAL REGION

by

HARALD SCHREIBER

DR. W. JUNK B.V., PUBLISHERS, THE HAGUE-BOSTON 1978

ISBN-13:978-94-009-9962-6   e-ISBN-13:978-94-009-9960-2
DOI: 10.1007/978-94-009-9960-2

Dissertation for the degree of a Philosophiae Doctor of the Philosophical Faculty of the
University of the Saarland, Saarbrücken.
Chairman of Faculty   : Prof. Dr. P. Steinmetz
Experts   : Prof. Dr. J. Schmithüsen
                          Prof. Dr. P. Müller
Date of last examination: April 5, 1973

Translated by Ingrid Schreiber, Saarbrücken
Softcover reprint of the hardcover 1st edition 1978

CONTENTS

VI

I. INTRODUCTION

The term 'dispersal centre' literally refers to the place from which a dispersal process started. However, it also implies the location of a centre of diversity, frequently even a recent one, as well as the location of a preservation centre or refuge during regressive phases in less remoted periods of earth's history, and that of a possible speciation centre. The latter may have been the place of differentiation of subspecies, or species, provided that the isolation period has been sufficiently long. The term deliberately brings the dynamics to prominence which is inherent in the process of dispersal. This dynamics is neither properly recognized when considering ranges 'whose recent structures more or less force us to consider them as static entities' (DE LATTIN 1967, p. 16)*, nor adequately taken into account by systematic typology.

Yet, it is, in fact, the change that takes place in a range, which is of great significance in evolutionary genetics. In the case of range regressions which may lead to a diminution or rupture of the range pattern according to the pressure of the endogenous or exogenous triggering factors, subsequently arising distribution obstacles may result in a geographical isolation of partial populations or population parts. Their sufficiently long spatial and reproductive isolation and the allele loss resulting from the change of the population size promote a divergent development. Due to the interrupted gene flow, mutation, recombination and selection have separate effects on the isolated parts of an original reproductive community (MAYR 1966).

The degree of speciation that was reached during refugial isolation is even increased, if a subsequent range expansion, the dispersal, takes place. This dispersal can then be expected to lead to an increase of the population from a relatively restricted number of individuals which had formed the gene pool till then. The conditions prevailing here are comparable to those described under the 'founder principle' (MAYR 1966, p. 529) in the case of colonizing species. Whether speciation has reached species level may chorologically be determined with certainty as soon as, in the course of dispersal, closely related forms have ceased to mix and occur sympatrically with mutual intersterility.

Apart from some exceptions, in which the reasons for an increase or decrease of the range of a taxon are easily visible, these processes elude our direct observation, especially when whole biota have been set in motion.

* 'die wir, notgedrungen, in ihrer rezenten Gestalt als statische Struktur zu erfassen versuchen'

A careful analysis of the present distribution pattern of animal or plant organisms, however, and a subsequent chorological investigation enable us to discover structures of an earlier distribution pattern beyond the establishment of recent ranges. This in turn provides insight into the dynamics of dispersal processes that have taken place since then. It is the purpose of this work to elaborate such structures in the Neotropical region on the basis of the distribution of the Sphingidae, a family of Lepidoptera Heterocera, and to interpret them by means of a comparative study of the species and subspecies involved.

The dispersal centres of the Holarctic region have been analysed by DE LATTIN (1957) on the basis of both vertebrates and invertebrates, especially Lepidoptera. As for South America, corresponding investigations have so far been made mainly by interpreting vertebrate ranges (MÜLLER 1970, 1973, HAFFER 1969). Only very recently have Heliconian forms been assigned to postulated refugia (BROWN 1975). However, further documentary evidence is required from other animal groups especially for a tropical continent, in which the processes of the glacial and pluvial times are less well discernible than on the north continents, and which appears to be much more complex faunistically due to the very fact that it distinguishes itself by a greater richness in species.

The Sphingidae have been chosen from the insects for the following reasons: their systematics are well known especially from the revision by ROTHSCHILD & JORDAN (1903); they have a predominantly tropical distribution; and their beauty and conspicuousness at all times earned them the particular interest of collectors, so that they are correspondingly well represented in museums of natural history.

It seemed important to discuss the nomenclature of the genera of the New World inasmuch as there have been modifications in recent years.

To include all of the 576 plotted range maps in the present work would be neither justified in terms of theme nor feasible in terms of volume. It is planned to publish the range maps as a separate atlas upon a more thorough revision and supplementation of the locality data.

I should like to thank the following institutions for their support in financing my stay in the United States and the journeys that had to be undertaken in connection with my investigations: the 'Deutsche Forschungsgemeinschaft', the Scholarship Scheme according to the 'Honnefer Modell', the 'Freunde und Förderer der Universität des Saarlandes' and the Carnegie Museum, Pittsburgh.

I owe great thanks to my respected teachers, the late Prof. Dr. G. DE LATTIN (Saarbrücken), at whose suggestion I undertook the present thesis, Prof. Dr. J. SCHMITHÜSEN (Saarbrücken), who was kind enough to accept me as doctorand upon the death of my Supervisor, and Prof. Dr. P. MÜLLER (Saarbrücken), who supervised the progressing of the present work in his capacity as mentor.

I am also deeply grateful to Prof. Dr. E. MAYR (Cambridge, Mass., USA) for arranging my stay at Harvard University and at the Museum of Comparative Zoology, and for his critical remarks on this work.

I am likewise very grateful to Mr. A.H. HAYES (BMNH, London) for his kindness to read over the English translation of my thesis.

Furthermore, I should like to thank the following gentlemen for permitting me to peruse their collections or collections they are in charge of, or granting me further help: A.H. HAYES, A. WATSON and Dr. I.W.B. NYE (British Museum of Natural History, London), Dr. W. FORSTER and Dr. W. DIERL (Zoologische Sammlung des Bayerischen Staates, Munich), Dr. H. SCHRÖDER (Senckenberg Museum, Frankfort), Prof. Dr. H. WEIDNER (Zoologisches Staatsinstitut und Zoologisches Museum, Hamburg), Dr. P. VIETTE (Muséum National d'Histoire Naturelle, Paris), Prof. Dr. P.J. DARLINGTON and Dr. J. BURNS (Museum of Comparative Zoology, Cambridge, Mass., USA), H.K. CLENCH and Dr. G. WALLACE (Carnegie Museum, Pittsburgh, Pa., USA), Dr. R.W. HODGES and Dr. J.F.G. CLARKE (United States National Museum, Washington, D.C., USA), Dr. F.H. RINDGE (American Museum of Natural History, New York, N.Y., USA), Prof. Dr. C.L. REMINGTON (Peabody Museum of Yale University, New Haven, N.J., USA), V. ROTH (SW Research Station of the AMNH, Portal, Arizona, USA), J.P. DONAHUE (Los Angeles County Museum, Los Angeles, Calif., USA), Dr. F. WERNER (University Collection, Tucson, Arizona, USA), NN (Arizona State University, Tempe, Arizona, USA), R.O. KENDALL (private collection, San Antonio, Tex., USA) and E.L. BRAUN (private collection, Friedrichsthal/Saar, Germany).

I should also like to thank Dr. J. McCOY (Pittsburgh, USA) who accompanied me on a excursion through Mexico, as well as Dr. W.F. REINIG (Hardt bei Nürtingen, Germany), Dr. R.H. CARCASSON (Vancouver, Canada) and Prof. Dr. F. VUILLEUMIER (New York, USA) who gave me the opportunity to discuss problems of this work.

I am also grateful to Dr. and Mrs. E. BÖHLER (Saarbrücken) for their help in plotting the localities on the range maps.

Finally I should like to tender my thanks to my wife INGRID for her understanding and manifold help as well as for translating this work into English.

3

II. METHODS

a. The importance of chorological investigations for systematics

For a long time, systematics maintained a leading position in biology. This is different today. Quite a number of biological partial disciplines are competing with it, such as physiology, genetics, biochemistry and biophysics, ethology, ecology and biogeography.

Part of the reasons for this change might lie in the fact that systematics has frequently been seen as a goal in itself; on the other hand, its results, even though they find their expression in a hierarchic system, are more difficult to verify than with the so-called exact sciences and the sub-domains of biology which tend to closely lean to these sciences. For each individual animal group specific methods of investigation have to be developed which are rarely transferrable to other groups. The possibility to evaluate the results in detail is, therefore, left to a small number of specialists in each case.

Even though, theoretically, the biological species concept is well recognized, in practical taxonomy specialists mostly depend on working with morphological methods, however differentiated they might be in that, with Lepidoptera for instance, they include microscopic examinations of the genitalia, and with vertebrates they are based on the evaluation of a variety of biometric data. It is true that a statistical corroboration can reduce the percentage of errors in investigations covering large series, or that the recourse to characters other than morphological ones — anatomical, enzymatic and other physiochemical characters — or to ecological, chorological and ethological criteria can also help to this end, as can the crossing test, even though, apart from being time-consuming and sometimes even unpracticable, it is not unequivocal, as it neglects the demand for 'natural conditions'.

Nevertheless, a certain degree of subjectivity cannot be avoided, when a specialist tries to weight the phylogenetic power of evidence which a character has in the sense of HENNIG (1950, 1966, 1969). MAYR (1969a, p. 219) writes on this point: 'Indeed, it is precisely this problem of how to determine what is a good character, a character with high weight, that has been a main source of concern among taxonomists', or 'the only reason why high weight is given to certain characters is that generations of taxonomists have found these characters reliable in permitting predictions as to association with other characters and as to the assignment of previously unknown species'. The difficulty of 'weighting', i.e. of deciding whether a character is plesio- or apomorphous or whether two taxa show

4

synapomorphies is complicated by the fact that convergencies which must not be adduced for relationship evidence result from adaptation and annidation experiencing a relative adjustment in equal or similar biomes, depending on the environmental conditions prevailing there. Furthermore, in practice, individual differences are frequently mistaken for good characters, a danger which is, of course, particularly great whereever species or subspecies are established on few or one single individual.

These observations emphasize the great importance inherent in good alternative methods in the attempt to leave behind the previously artificial system and gradually approach a natural one.

Studies of the geographical distribution are particularly well suited for checking merely morphologically ascertained diagnoses (cf. MAYR 1969a, p. 140: 'Geographical characters are among the most useful tools for clarifying a confused taxonomic picture and for testing taxonomic hypotheses'; cf. also SIMPSON 1965, NICHOLS 1962 and HENNIG 1950, 1969).

Range analyses are also of particular importance, because in contrast to morphological investigations which consider individuals only, they deal with populations. Just like an ecological observation, the distribution pattern is a bionomic criterion, since the distribution is the chorological expression of the potential living demands.

The distribution pattern gives a clue to the ecological demands an organism makes on its environment. Vice versa, animals and plants who ecological valencies are known, become indicators to areas with equal or similar environmental conditions through their distribution ranges (MÜLLER & SCHREIBER 1972).

Besides the recent-ecological correlation existing between range and environment, the distribution pattern has a historic-genetic information content, whereby it is true that 'a large part of what we call history today represents the ecology of previous times' (DE LATTIN 1967, p. 20)*.

The method applied in the present investigation strives particularly at following up the factors which played a decisive role in speciation and range formation during the more recent periods of earth's history. In this connection, DE CANDOLLE (1855) expressed the view quoted and endorsed by REINIG (1937) that 'everywhere in the distribution of living beings the previous causes predominate over the conditions of the present state'.**

b. Special method applied

The method applied in this work and the concept on which it is based are a synthesis of the conceptions developed in the past and altered or added views

---

* 'das, was wir heute als Historie bezeichnen, zu einem beträchtlichen Teil die Ökologie früherer Zeiten repräsentiert'.
** 'Überall herrschen in der Verbreitung der Lebewesen die früheren Ursachen vor über die Bedingungen des gegenwärtigen Zustandes'.

developed from the critical concern with these conceptions.

HOFMANN (1873) used isopores (lines connecting points of equal species frequencies) to mark the species fall between the common place of origin of species, such as the glacial refuge, and the border of the postglacial invasion area.

VAVILOV (1926, 1931) tracing back cultivated plants to their places of origin, found areas of great species diversity which he called gene centres*.

REINIG (1938) found that, besides refuges, the glacial areas of retreat were also centres of species diversity and centres of postglacial dispersal.

It is owing to DE LATTIN (1957, 1967) that by way of a consistent application of the comparable-chorological method an ecologic-phylogenetic causal analysis of distribution centres has been supplied which is necessary to permit an objective conclusion as to centres of dispersal.

A basic prerequisite to this is the starting from ranges of 'real units' (DE LATTIN 1967), as represented especially by species as compared to higher taxa. By first excluding deliberately polytypic species and species with large ranges from the consideration and having a look at the ranges of all remaining species plotted on a map of the area under investigation, one will come across selected areas whose species share is uncomparably higher than that of the surrounding areas. These centres of distribution which REINIG (1950) called 'nuclear areas' will, however, only unreveal themselves as dispersal centres, if they can also be identified as the subspeciation centres of polytypic species which, according to the allopatric distribution of their subspecies (cf. MAYR, LINSLEY & USINGER 1953), must have a polycentric distribution.

In addition, a comparative consideration of stationary and expansive types must be undertaken.

DE LATTIN (1957, 1967) called the aggregate of all species and subspecies belonging to a dispersal centre a faunal circle and its species and subspecies its faunal elements according to a term used by REBEL (1931).

On the other hand, species with a large distribution range, which are polycentric and at the same time primarily or secondarily monotypic cannot, or at least not with certainty, be related to a dispersal centre. By analogy with the terminology used by HENNIG (1950, 1966, 1969) MÜLLER (1972) introduced the term 'apochor' for species whose ranges experienced an expansion far off the dispersal centre, whereas the term 'plesiochor' is used for monotypic species or subspecies which are still in connection with only one − their − dispersal centre and which alone (DE LATTIN 1967) can be regarded as the basic elements of zoogeographical working.

As for the subspecies which are, by necessity, allopatrically distributed, the centre of dispersal is also their centre of origin. The fact that their differentiation required isolation leads to the assumption of a regression phase, during which the environmental conditions required by the ecological valency of the respective taxa

---

* in the sense of allele centres (REINIG 1938, DE LATTIN 1967)

have undergone a restriction; with that a relative stability of the ecological valency is taken for granted.

Every species, too (cf. MÜLLER 1971, p. 2), 'possesses at least one dispersal centre which represents a structure homologous to the centre of origin. During the evolution of a taxon, however, both structures can become widely separated from each other. The analysed dispersal centre, therefore, only permits conclusions regarding the latest regressive range phase that was important for the respective taxa'.*

In contradistinction to the rather arbitrary method so far in use to subdivide zoogeographical regions into subregions, provinces, districts etc. on the basis of anthropomorphic conceptions, the faunal circles identified by real systematic units permit a subdivision which is equally applicable to all animal groups, no matter how different they are (cf. DE LATTIN 1957, 1967, MÜLLER 1970, 1971, 1973).

An analysis of the relationship of the established dispersal centres will finally disclose the fact that the relationship is, on the one hand, a function of the phylogeny of their faunal elements (MÜLLER 1971), especially the polytypic species and, on the other, a function of the ecological conditions prevailing in the centres.

The degree of the ecologically conditioned relationship can be determined by classifying the centres according to their arboreal, eremial or oreotundral character (REINIG 1950, DE LATTIN 1957) or by splitting them further up according to the specific biomes to which they belong.

DE LATTIN'S statement (1967, p. 356) that 'dispersal centres and faunal elements must be generally distributed basic zoogeographical structures neither bound to the Holarctic region with regard to their location nor to the Pleistocene with regard to their origin'*, could be unequivocally confirmed by MÜLLER (1970, 1973) with an investigation of terrestrial vetrebrates. The results of his studies are further confirmed by the respective entomological studies carried through in the present work.

c. Procedure and cartographic presentation

In elaborating the present work the following zoogeographical and systematic literature has been especially helpful: BEEBE & FLEMING 1945, BIEZANKO & ZOPP 1954, CARY 1951, 1963, CLARK 1917-1931, DANIEL 1949a, b, DAR-

---

* ... 'besitzt mindestens ein Ausbreitungszentrum, das eine homologe Struktur zum Entstehungszentrum darstellt. Im Verlauf der Evolution eines Taxon können sich jedoch beide Strukturen erheblich voneinander entfernen. Das analysierte Ausbreitungszentrum läßt also nur Rückschlüsse auf die jüngste regressive Arealphase, die für die betreffenden Taxa bedeutungsvoll war, zu'.

* '... Ausbreitungszentren und Faunenelemente generell verbreitete zoogeographische Grundstrukturen sein müssen, die lagemäßig weder an die Holarktis noch entstehungsmäßig an das Pleistozän gebunden sind'.

LINGTON 1957, DE LATTIN 1957, 1967, DRAUDT 1931, HODGES 1971, HOFFMANN, C.C. 1942, HOFFMANN, F. 1934, KAYE & LAMONT 1927, KERNBACH 1952-1968, LICHY 1943-1968, MAYR 1969a, b, MOOSER 1940, 1942, MOSS 1912, 1920, MÜLLER 1970, 1971, 1973, OITICICA 1939, 1942, 1946, ORFILA 1933, REINIG 1937, 1938, 1950, ROTHSCHILD & JORDAN 1903, 1910, 1916, SCHILDER 1956, SCHREITER 1926, UDVARDY 1969, URETA & DONOSA 1956, WALLACE 1876.

The first step towards obtaining range maps that could be interpreted zoogeographically was to collect and map verified locality data.

It has only recently been apparent how important it is to cooperate on supra-regional and international levels in collecting locality data of plant and animal species and to record them in central places, as has been demanded by DE LATTIN (1967) or, at an earlier date, by NETOLITZKY (1939). By availing oneself of the modern aids of computer technology this has meanwhile become easier even with the invertebrate groups (cf. HEATH & LECLERCQ 1970) which are, for the most part, very rich in species.

Locality data can be obtained either from the existing literature or by collecting in the areas which are the subjects of investigation, or, finally, by examining the material contained in museum collections.

Information in literature is too sporadic and often not accurate enough. No doubt, collecting in the field is the best method, but it would not have yielded working data within measurable time. For this reason, I examined and evaluated the major sphingid collections of the New World which are to be found primarily in Great Britain and the United States. The British Museum (Natural History), London, possesses the ROTHSCHILD Collection, and the Carnegie Museum, Pittsburgh, Pa., USA, the collections of several lepidopterists as those of CLARK and HOLLAND. CLARK was an impassioned sphingid collector whose aim was to make of his collection the most important future reference collection of spingids at least of the western hemisphere (CLARK 1922).

I think it is only fair at this point to appreciate the initiative of a private collector who spared no effort in pursuing his aim to make his contribution, as he says in great modesty (CLARK 1923), 'in one tiny corner of the world of nature', thus helping to lay the grounds for subsequent investigations like the present.

The following is a list of the public museums and private collections which I perused; the sequence of enumeration follows the quantitative importance of the collections in terms of their sphingid material from the New World:*

---

* The figures in brackets represent the number of recorded locality data.

BMNH	British Museum (Natural History), London, England	(22,815)
CM	Carnegie Museum, Pittsburgh, Pa., USA	(18,500)
USNM	United States National Museum, Washington, D.C., USA	(8,101)
AMNH	American Museum of Natural History, New York, N.Y., USA	(5,455)
PMY	Peabody Museum of Yale University, New Haven, Conn., USA (including the collection of M.M. CARY)	(4,604)
ZSM	Zoologische Sammlung des Bayerischen Staates, München, BRD	(4,400)
MCZ	Museum of Comparative Zoology, Cambridge, Mass., USA	(3,002)
ZIMH	Zoologisches Staatsinstitut und Zoologisches Museum, Hamburg, BRD (including the collection of K. KERNBACH)	(1,787)
LACM	Los Angeles County Museum, Los Angeles, Calif., USA	(1,726)
SMF	Senckenberg Museum, Frankfurt, BRD	(1,031)
MHNP	Muséum National d'Histoire Naturelle, Paris, France	(1,019)
BGSS	Biogeographische Sammlung, Saarbrücken	(881)
ASUC	Arizona State University Collection, Tempe, Arizona, USA	(343)
PSB	Private Collection LUDWIG BRAUN, Friedrichsthal/Saar, BRD	(318)
UCT	University Collection, Tucson, Arizona, USA	(310)
PSK	Private Collection ROY O. KENDALL, San Antonio, Texas, USA	(126)
SWRS	South West Research Station of the AMNH, Portal, Arizona, USA	(41)

The locality data registered this way (totalling 74, 459) were supplemented by locality data from literature as well as data obtained from my own collecting in the border zone between the Nearctic and the Neotropical regions.

In mapping the locality data a quantitative inventory has been taken by the use of different symbols, with the qualification, however, that only single observations are absolutely apparent from the maps, whereas all other symbols represent observations from 2 to 10, 11 to 30 and more than 30 specimens.

For the presentation of centres of distribution, in which several individual ranges have to be plotted on a map of the region, contour graphs are particularly well suited and can be combined with different superficial markings or range graphs marked with dots (cf. MÜLLER 1970, 1971, 1973).

Plotting all possible ranges on the respective map of the region would certainly have jeopardized the clearness of illustration. For this reason, selected ranges have been represented cartographically, whereas the remaining species or subspecies are listed in the underline.

In order to fulfil the double demand for an optimum illustrative value of the maps and a maximum power of evidence, the grid as given by the network of longitudes and latitudes has been provided with the numbers and letters of the coordinate grid system used in atlasses for locating geographical places.

A list of the more than 1000 localities – recurring with the different species according to the position of their ranges – is accompanied by a reference guide-map,

9

in which these localities have been entered and numbered. With the aid of these numbers and the respective grid references as given in the index, each individual dot on the range maps marking a locality can be determined.

In drawing up the range maps the TIMES Atlas (1967) served as the basic reference book. Where older locality names were involved which might have been difficult to locate in recent works of reference due to renaming, the 'Maps of Hispanic America' (1942-1944) with the pertaining index have been used. In addition to this, the 'Dicionário GEOGRÁFICO Brasileiro' (1967) has been used. As for Central America, the locality list given by SELANDER & VAURIE (1962) was particularly helpful.

As a help for determination a slide collection has been compiled, and, wherever types have been available, pictures were taken; for the rest, pictures were taken of a typically looking representative of the species, and, with polymorphic species, of several specimens showing the extremes of variation.

III. DELIMITATION OF THE AREA UNDER INVESTIGATION AND OF THE PERIOD OF EARTH'S HISTORY ON WHICH STATEMENTS CAN BE MADE ON THE BASIS OF AN ANALYSIS OF DISPERSAL CENTRES

The area investigated in this work is the Neotropical region as it has been defined as early as in 1858 by SCLATER as a zoogeographical region and confirmed in 1876 by WALLACE. Many authors (WALLACE 1876, SCHILDER 1956 and UDVARDY 1969) such as MURRAY (1866), HUXLEY (1868), BLYTH (1871), ALLEN (1871) and others have attempted to redefine this area, it is true, yet the original classification is basically still valid even today (cf. DE LATTIN 1967).

There have also been recent attacks launched against a regional concept in that one tends to propagate a special subdivision according to ecological considerations. I would like to profess the regional concept in the sense of DARLINGTON (1957, p. 422), who says: 'The system of faunal regions, then, represents the average, gross pattern of distribution of many different animals with more or less different distributions', a division mainly based on practical considerations.

Vegetation geographers (ENGLER 1882, WALTER 1954, GOOD 1964) exclude the southern Chilian part from the Neotropical floral region, whereas zoogeographers rather tend to argue about the northern delimitation of the Neotropical region. Depending on the respective animal group which is paramount in the considerations, they postulate the inclusion of or an exceptional status for the West Indian Islands, and for the continent either a clearly defined faunal border or a more or less wide transition zone (cf. SCHMIDT 1954).

A faunal border between the Neartic and the Neotropical regions which can be verified by the range borders of many of the investigated species, runs within Mexico (cf. WALLACE 1876, SIMPSON 1953 and others) from the Rio Grande on the east coast to the area north of Mazatlan in the west and excludes the central elevated plateau bordered by the Sierra Madre Oriental and the Sierra Madre Occidental. On the plateau tapering in the south, Nearctic elements penetrate far into the south. A reasonable border with regard to the present investigation seems to be the Balsas depression, the area south of which is also referred to as Central America (NEEF 1968). WALLACE (1876) thought that the open highlands of South Mexico and Guatemala should perhaps also be included in the Nearctic region, a consideration which is, however, not reflected in his cartographical presentation of the zoogeographical regions in the same work.

The species whose range borders are shown in fig. 1 are species which, in spite of their great ecological valency and vagility, are not able to overcome the barriers existing in this area; the ranges of the Neotropical species spread on either side of the uplands, whereas the Nearctic species claim a sphenoid extension of their ranges on the plateau.

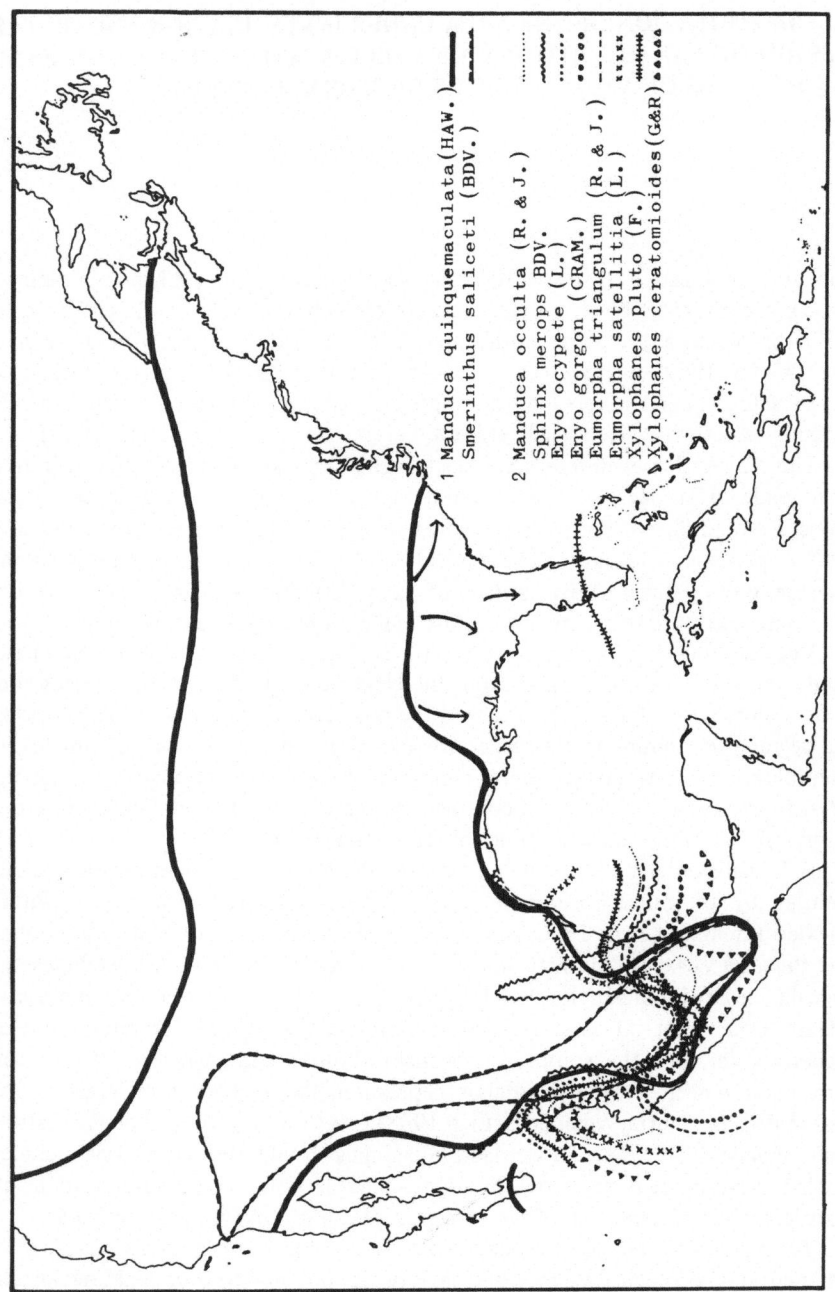

Fig. 1. Documentation of the faunal border between the Nearctic and the Neotropical regions on the example of widely distributed sphingid species: 1. ranges of Nearctic species with regard to their southern limits; 2. northern limits of wide-spread Neotropical species.

Besides the limit as predetermined by the relief, arid conditions have an additional hampering effect on distribution in the northwest of Mexico.

DARLINGTON (1957), in the introduction to 'Zoogeography', emphasizes that the correct understanding of zoogeographical correlations depends largely on the choice of an appropriate mapping material.

It appears just as important that zoogeographers mark off the spaces of time about which they want to make a statement, and, above all, that they see them in their correct temporal relation. This is of particular importance when working with animal groups in which there is no fossil material to revert to. It is exactly these groups, among them the Lepidoptera, where, like HENNIG (1969) says, a careful analysis of the geographical distribution is of paramount significance, if one wants to clarify phylogenetic questions.

A proper understanding of the temporal proportions, however, is just as important, if one wants to date established phylogenetic correlations, or if one tries to draw conclusions from the known ecological valency of certain taxa on the palaeoclimate and the genesis of landscapes.

In fig. 2 it has been tried to illustrate the different geological periods in their correct temporal relation by using a 1 mm stretch for a period of 10,000 years.

The periods since the assumed origin of the earth 4,000 millions of years ago until the beginning of the Cenozoic are represented by a block that corresponds to a stretch of as much as 400 m. The thread of time taken up at the beginning of the Tertiary would then be as long as 7 m for the whole Tertiary, of which 1.20 m would fall to the share of the Pliocene. The Pleistocene (the changes between cold and warm periods are represented by a curve) would correspond to a stretch of 20 cm, whereas the Postglacial would measure 1 mm only. (A division of the Postglacial in South America into dry and wet phases is given by MÜLLER & SCHMITHÜSEN 1970).

In studying the distribution of animals and plants and in searching for an explanation of their present distribution patterns as well as in assessing phylogenetic correlations one should always be aware of the fact that only the uppermost level at the top of the 1 mm long stretch is accessible. For all the rest we have to rely on fossil findings which are scarce in comparison with the periods of time, or which may even be lacking completely for lack of preservable hard body parts, such as inner or outer skeletons, or due to the particularly quick decomposition of organic material in tropical areas.

The efforts of phylogeneticists to reconstruct, as exactly as possible, the phylogenetic changes on the basis of the existing fossil and recent material, can be supported by evidence available to zoogeographers alone, provided that the methods they use are exact and come up to the demand to be verifiable and reapplicable.

The concept of dispersal centres according to DE LATTIN (1957) can claim to fulfil this demand, since, as shown through the investigations carried through so far with different animal groups (MÜLLER 1970, 1973, ANT 1964, GROSS

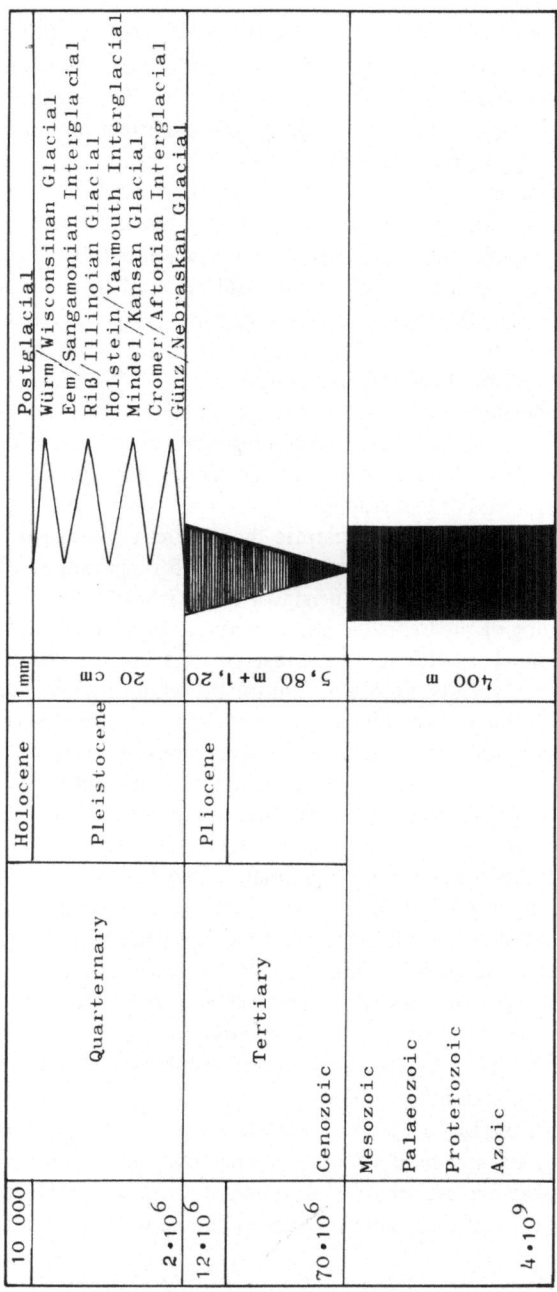

Fig. 2. Periods of earth's history in temporal relation to each other: 1 mm is equivalent to 10 000 years; time given in million years since beginning of period

14

1961, 1962, WAGNER 1961, ROESLER 1965, KEPKA 1969) there has been a basic conformity of the results.

It is of primary importance in this connection not to search for the place of origin of high taxa, as with a higher taxon the probability that the centre of origin and the centre of dispersal have become separated from each other increases in proportion to time (MÜLLER 1970, 1973).

The attempt to find out about the most recent regressive phase so important for the latest speciation process represents an effort to understand phylogeny from the only point accessible through the distribution of real units, as represented by species and subspecies (DE LATTIN 1967).

A comparison of the recent distribution with the previous distribution pattern as it has been existing at the time of the latest regressive phase that was paramount and decisive for the respective organisms, and as it is suggested by the position of the refuge areas, enables us, for the first time, to procede to a meaningful subdivision of a faunistic region which throws light on the origin of the ranges as they exist today.

The regional subdivisions by previous authors, some of which are listed in table 1 (WALLACE 1876, LYDEKKER 1896, SCHILDER 1956), are mainly based on the given topographic and climatological facts and the pattern of certain vegetation units respectively; this, however, only considers the recent-ecological component. Other authors (KRÜGER 1932, CABRERA & YEPES 1940, 1947 and MELLO-LEITÃO 1946 who mainly considered phytophagous animals) tried to fix their subdivision of the Neotropical region on the fauna, but these subdivisions did not reach the level of general validity, because the methods used have been too varied, or they applied only to a specific animal group in each case. UDVARDY (1975) subdivided the Neotropical region into 47 biogeographic provinces (30 on the mainland), but, as he says, he paid less attention to the historic element than to vegetational entities as shown by HUECK (1972) and others.

The results of previous investigations coincide to a certain degree with those of the analysis of dispersal centres in those cases in which the individuality of a region can justly be inferred from its share of endemics. These investigations have mainly been carried through with islands which, at all times, have been privileged objects of zoogeographical studies, because they show the isolation conditions necessary for speciation and because they are clearly delimited areas.

That this is also true of sphingids is reflected by the high endemism on the Galápagos Islands, as is obvious from the works of CLARK (1926), KERNBACH (1962b, 1964) and HAYES (1975). The West Indian Islands, too, which have been investigated by CARY (1951), are very rich in endemics.

The method employable for islands to obtain a rough natural grouping by way of determining their share of endemics breaks down already in Central America, to say nothing of the South American Continent. The share of endemics in the fauna as a whole can be determined, it is true, but then there is a lack of sufficient criteria for outlining specific areas, with which the largest possible number of animal groups are correlated. At this point, the comparative chorological method can help.

Previous zoogeographers used to make statements on far remoted epochs when trying to adduce evidence for hypotheses that had been made on the part of geologists. As compared to this, the step which can be made back into the past by means of the method of dispersal centres is relatively small. It leads to the latest differentiation phase of each of the faunal elements of a centre (MÜLLER 1970, 1971, 1973), which may have taken place as recently as in the Postglacial or the latest phase of the Glacial.

Table 1. Previous attempts to subdivide the Neotropical region.
(KRÜGER's suggestion is based on investigations with butterflies.)

WALLACE (1876)	LYDEKKER (1896)	
South Temperate America	Andine Subregion	
or the Chilian Sub-region	Pampean Subregion	
Tropical South America,	Amazonian Subregion	
or the Brazilian Sub-region		
Tropical North America,	Astecan Subregion	
or the Mexican Sub-region	Maian Subregion	
The West Indian Islands,	Antillean Subregion	
or the Antillean Sub-region		
KRÜGER (1932)	SCHILDER (1956)	
Guayana Zentrum	antillisch	
SO Brasil. Zentrum	dendrogäisch	centralamerikanisch
Oberamazon. Zentrum		brasilianisch
Pazifisches Zentrum	andinisch	peruanisch
S. Mittelam. Zentrum		patagonisch
CABRERA & YEPES (1940)	MELLO-LEITÃO (1946)	
Tropical	Brasiliana	Guianense
Amazonico		Hiléia
Sabânico		Cariri
Sub-Tropical		Guarani
Tupi	Andino-Patagônica	Tupi
Incásico		Incásica
Andino		Chilena
Chileno	Centro-Americana	Subandino Pampásico
Subandino		Yucateca
Pampásico		Guatemalteca
Patagônico	Antilhense	Dariênica/Istmica
		Cubana
		Jamaicense
		ʾispaniólica
		ʃ ·toricense
		Micro Antilhense

IV. SYSTEMATICS

a. Sphingidae LATR.

The name *'Sphinx'* (cf. JORDAN 1911) goes back to REAUMUR (1736) who used it for *Sphinx ligustri*. It refers to the habit of the caterpillars of many sphingid species to assume a posture, if disturbed, which resembles that of the Egyptian legendary beast. LINNE then adopted *Sphinx* as an appropriate genus name.

There is not so far unanimity on the phylogenetic position of the Sphingidae within the Lepidoptera (cf. table 2).

In looking for criteria which will be equally valid in the description of any of the sphingid species it is necessary to generalize for the sake of clearness, even though this method fails to do justice to exceptions that can be found to almost any phenomenon cited as a rule. This is true, for instance, in describing the size of the moths, the shape of their wings or their antennae, the size of their eyes, the length of the proboscis or their great flying ability so frequently emphasized (cf. SEITZ 1931, FORSTER & WOHLFAHRT 1956).

A generalizing conclusion drawn from this latter phenomenon would not come up to the facts. DIERL (1970) writes about this: 'The hawk-moths (Sphingidae) are particularly well suited for an ecological-zoogeographical investigation in spite of the fact that, powerful flyers as they are, they show vast distribution ranges whose borders could be expected to be indistinct'.*

The Sphingidae belong to the Macrolepidoptera Heterocera. Within the Neotropical genus *Cocytius* it is especially the females of the species *C. antaeus* which reach span widths up to almost 20 cm (fig. 3). Species of small sizes are the exceptions. The smallest species (JORDAN 1911) is the Madagascan *Sphingonaepiopsis obscurus* whose forewings are only 1 cm long. In the New World, too, there are species of the genera *Cautethia* (fig. 3) and *Euproserpinus* (fig. 3) which, with a forewing length of hardly 1.5 cm, are not much larger.

The strong body is described as being cigar-, cone- or streamline-shaped. It ends in a long and slender abdomen, especially in case of the exclusively American *Xylophanes* species.

Apart from some exceptions, the forewings are long and narrow. The wings

---

* 'Die Schärmer (Sphingidae) eignen sich für eine ökologisch-tiergeographische Studie ganz besonders, obwohl sie als kraftvolle Flieger eigentlich große Verbreitungsgebiete aufweisen und ihre Grenzen verwischen sollten'.

17

may be connected by the frenulum on the hindwing and its counterpart on the forewing, the retinaculum (fig. 4.7). In the case of some Smerinthini, the frenulum is reduced. In the female, the frenulum consists of several small bristles, the retinaculum is lacking. The coupling device makes the wing beat more efficient.

According to AMANS (1883, cf. SEITZ 1927), wings like those of the Sphingidae are simply ideal flying organs. They enable the Sphingidae to make 25 to 45 wing beats/second (DAVIDSON 1965), which lies comparatively little under the beat frequency of humming birds of 50 to 78 wing strokes/second (MEISE 1969).

In other respects, too, the sphingids promt a comparison with humming birds. They, too, are found hovering before deep-throated flowers taking nectar without

Table 2. Different classifications of Lepidoptera 1. KAESTNER (1973), 2. CARCASSON (1968a according to IMMS 1957). 3. FORSTER (1954).

1.	3.	UDVARDY (1975) Biogeographic Provinces
Papilionoidea	Noctuoidea	1. Campechean
Hesperioidea	Notodontoidea	2. Panamanian
Sphingoidea	Zygaenoidea	3. Colombian Coastal
Bombycoidea	Calliduloidea	4. Guyanan
Noctuoidea	Cochlidioidea	5. Amazonian
Geometroidea	Sphingoidea	6. Madeiran
Pyralidoidea	Drepanoidea	7. Serra do Mar
Zygaenoidea	Uranioidea	8. Brazilian Rain Forest
Tortricoidea	Geometroidea	9. Brazilian Planalto
Tineoidea	Papilionoidea	10. Valdivian Forest
Cossoidea	Saturnioidea	11. Chilean Nothofagus
Incurvarioidea	Bombycoidea	12. Everglades
Stigmelloidea	Endromidoidea	13. Sinaloan
Hepialoidea	Pyralidoidea	14. Guerreran
Eriocranioidea	Hesperioidea	15. Yucatecan
Mycropterygoidea	Pterophoroidea	16. Central American
	Gelechoidea	17. Venezuelan Dry Forest
	Copromorphoidea	18. Venezuelan Deciduous Forest
	Hyponomeutoidea	19. Equadorian Dry Forest
	Castnioidea	20. Caatinga
	Tortricoidea	21. Gran Chaco
	Psychoidea	22. Chilean Araucaria Forest
	Glyphypterygoidea	23. Chilean Sclerophyll
	Plutelloidea	24. Pacific Desert
	Tineoidea	25. Monte
	Cossioidea	26. Patagonian
	Incurvarioidea	27. Llanos
	Nepticuloidea	28. Campos Limpos
	Hepialidae	29. Babacu
	Eriocn iidae	30. Campos Cerrados

2.

Papilionoidea ─┐
Hesperioidea ─┤ ┌─ Geometroidea
Castnioidea ──┤ ├─ Noctuoidea
Pyralidoidea ─┘ ├─ Sphingoidea
                 ├─ Bombycoidea
                 ├─ Calliduloidea
                 ├─ Cossoidea
                 └─ Psychoidea
Tineoidea ────
Tortricoidea ──
      DITRYSIA
Hepialoidea ──┐┌── Incurvarioidea
Eriocranioidea┘└── Stigmelloidea
      MONOTRYSIA
          |
      ZEUGLOPTERA

31. Argentinian Campos
32. Uruguayan Pampas
33. Northern Andean
34. Colombian Montane
35. Yungas
36. Puna
37. Southern Andean
38. Bahamas-Bermudan
39. Cuban
40. Greater Antillean
41. Lesser Antillean
42. Revilla Gigedo Island
43. Cocos Island
44. Galapagos Islands
45. Fernando de Noronja Island
46. South Trinidade Island
47. Lake Titicaca

Fig. 3. Selected sphingid species: 1 = *Cocytius antaeus medor* ♀; 2 = *Smerinthus cerisyi* ♂; 3 = *Amplypterus gannascus*; 4 = *Cautethia grotei*; 5 = *Cressonia juglandis* ♂; 6 = *Manduca sexta paphus*; 7 = *Euproserpinus phaeton*; 8 = *Manduca rustica*; 9 = *Enyo ocypete* ♂; 10 = *Xylophanes chiron nechus*; 11 = *Oryba kadeni*; 12 + 13 = *Erinnyis ello* ♂, ♀.

19

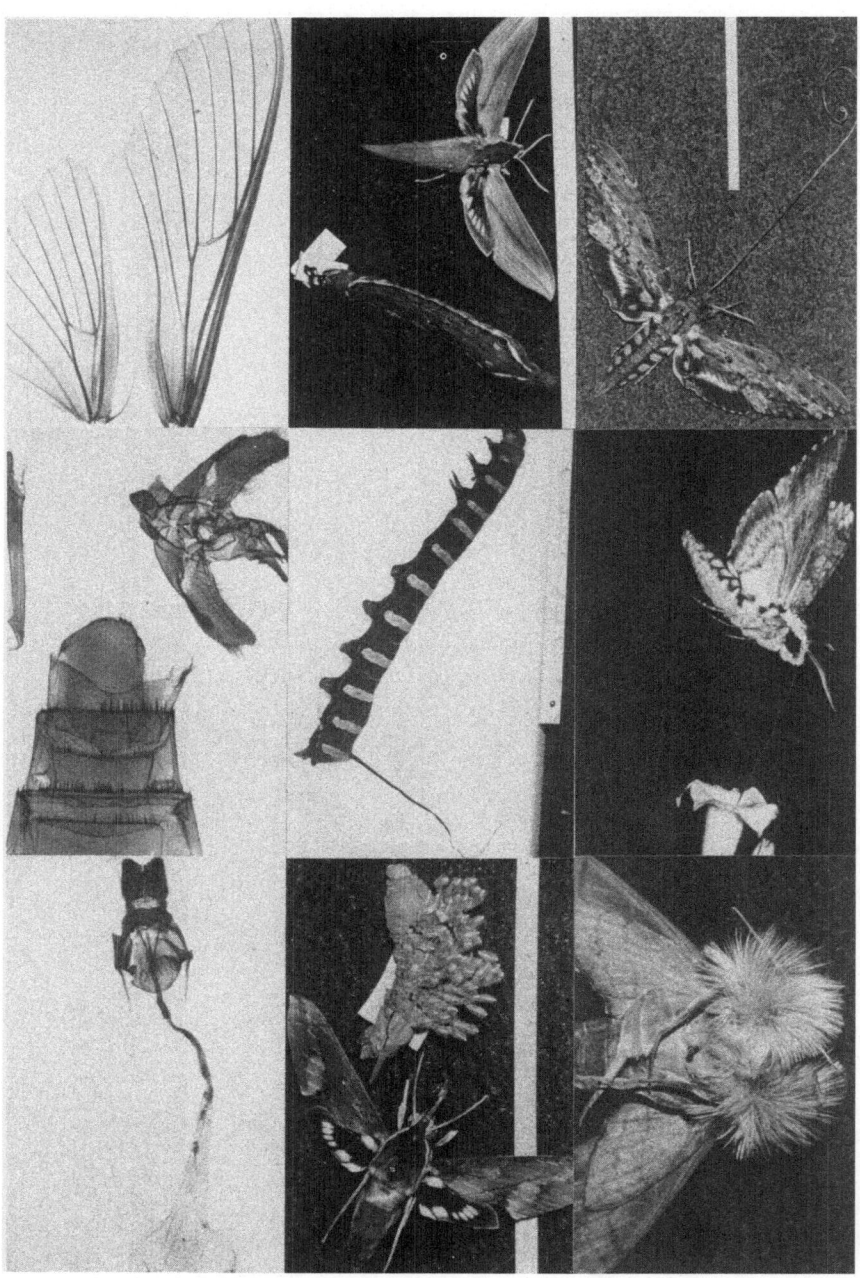

Fig. 4. Morphological details of Sphingidae: 4.1 *Neococytius cluentius*, proboscis unrolled; 4.2 *Manduca sexta* feeding on tobacco flower; 4.3 *L   orpha anchemola* showing scent organs on the fore-coxae; 4.4 *Xylophanes tersa*, imago anu ˍarva showing mimicry; 4.5 *Isognathus excelsior*, larva with long caudal horn; 4.6 *Xylophanes chiron nechus*, imago and larva, the latter infested by Hymenoptera; 4.7 *Erinnyis obscura*, wings of one side, scales removed; 4.8 *Erinnyis ello*, male genitalia. 4.9 *Neococytius cluentius*, female genitalia.

needing a support on the flower (fig. 4.2). They are able to do so due to their proboscis which, in some species, reaches an enormous length. HODGES (1971) mentions *Neococytius cluentius* as the sphingid with the longest tongue (255± mm) (fig. 4.1). According to AMSEL (1938) the Neotropical *Amphimoeca walkeri* is said to be the sphingid with the longest proboscis (up to 28 cm).

In contrast, there are species whose tongue is as much reduced as to be non-functional. Extreme examples are *Polyptychus* species from Africa (KERNBACH 1962a) whose tongues are reduced to stumps of 0.5 mm. It is interesting to look at the number of short- and long-tongued species within the Americas. From KERNBACH's investigations it results that, in North America, 7 of 33 species and subspecies have short tongues, whereas out of 206 species investigated in Central and South America *Orecta lycidas* and *Orecta lycidas eos* are the only ones to have short tongues of 1 cm length. (*Orecta fruhstorferi* and *Orecta acuminata* have not been investigated).

The Sphingidae can also be recognized by their large naked eyes and their prismatic, filiform or setiform antennae which are frequently clubbed and usually hook-shaped. The males of the Smerinthini have pectinate antennae. The antennae of the species of these genera show sexual dimorphism. In many sphingid species, such as *Erinnyis ello* (fig. 3) and *obscura, Enyo gorgon* and *ocypete* (fig. 3 – only the figured male has the light-yellow hindwing patches) or *Pseudosphinx tetrio*, sexual dimorphism manifests itself by different colour patterns or markings and, in addition, by the size of the normally larger females. In the males, the basis of the lower abdomen shows a scent gland located in ventrolateral pits, from which tufts of long hair may protrude. Some other species also have scent organs at the fore-coxae (fig. 4.3). In addition to this, in the *Enyo*-species *gorgon, taedium, bathus* and *cavifer*, the forewings of the males have a subcostal fold filled with scent scales and fleecy hair.

The stout legs (fig. 5) bear spurs at the tibiae and spines or rows of setae at the tarsi which lend the sphingids a good hold on any support even in stormy weather. SEITZ (1927) rather sees in these characteristics a protection against the danger of being eaten. The spines and spurs are so pointed that one may hurt oneself in handling the moths. A pulvillus may be present at the claw-segment of the tarsus and is used as a key characteristic; it is lacking, for instance, in the species of the genus *Sphinx*.

The venation (fig. 4.7) is very similar in all sphingids and can therefore not be used for taxonomical differentiation (ROTHSCHILD & JORDAN 1903, CAR-CASSON 1968).

Besides other morphological features, studies of the genitalia are especially helpful in such an approach (fig. 4.8, 4.9). They consist of a variety of chitinous parts which vary from species to species, but are stable wi in each species.*

There is a certain homogeneity of their outer appearar.. which makes it easy to determine the sphingids on the family level, but, on the other hand, com-

* The genital slides have been made by Mrs. ADAMS in the USNM, Washington.

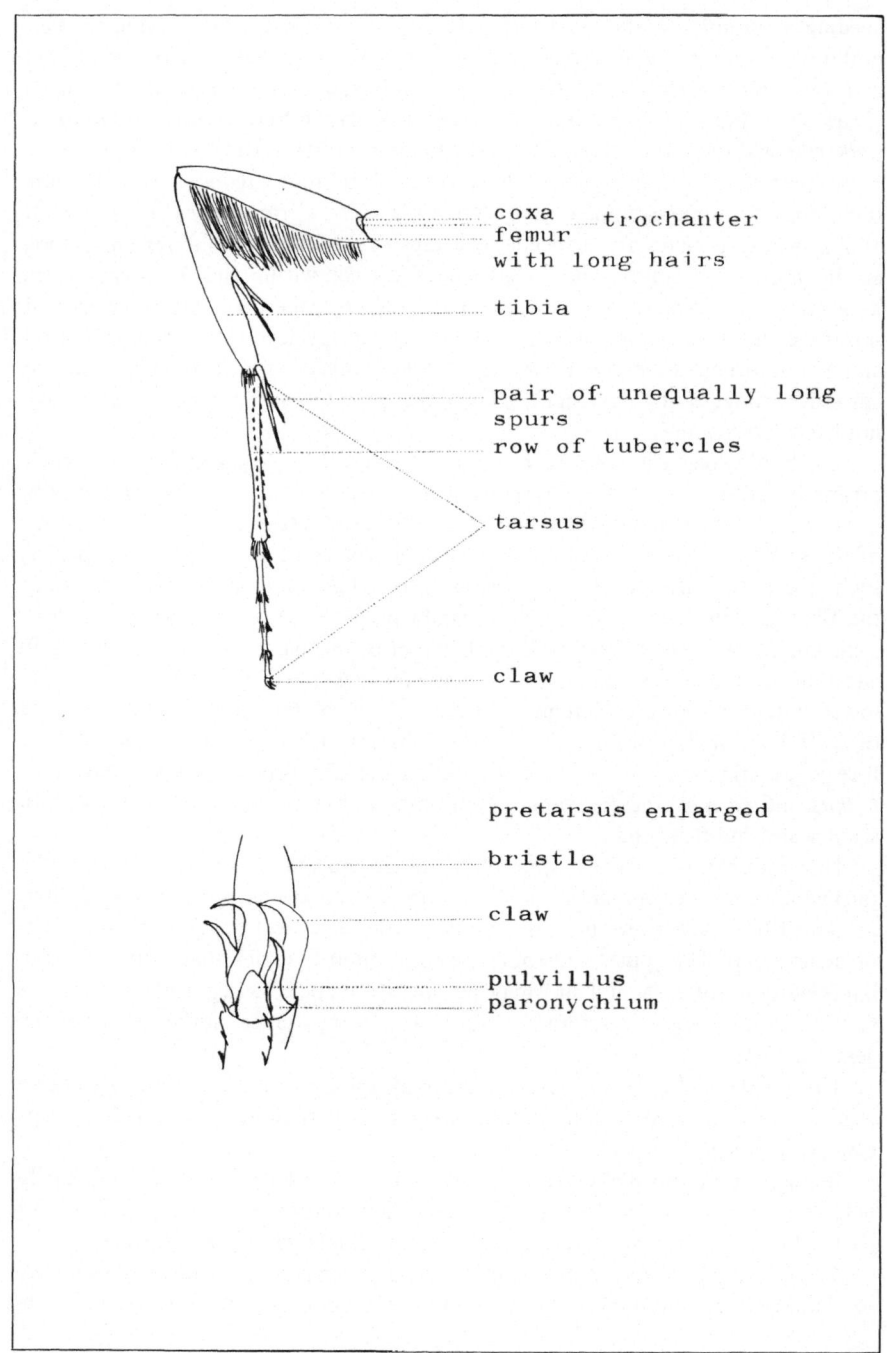

coxa ..... trochanter
femur
with long hairs

tibia

pair of unequally long
spurs
row of tubercles

tarsus

claw

pretarsus enlarged

bristle

claw

pulvillus
paronychium

Fig. 5. Manduca rustica, hindleg.

plicates the differentiation of sub-units (SEITZ 1931). Yet, the sphingids do have a variety of morphologically distinguishable characteristics, such as the shape of individual body parts or appendices, colour patterns and markings, which can be used for determination.

Whereas the shape of body and wings represents an adaptation connected with the flying habits, and the large compound eyes – especially conspicuous with *Oryba kadeni* (fig. 3) – can be explained by the predominantly crepuscular living habits, the long proboscis can be seen as an adaptation to the food plants of the imagines. This is particularly conspicuous in cases, where the visited flowers have deep calyces which open at the beginning of the flying period of the sphingids (SEITZ 1931), or if specific orchid flowers with long spurs are involved, at whose bottom the nectar collects. The insects visiting these flowers must have a long tongue with which they can reach the nectar through the flower, especially if their visit should also be of profit to the plant. With that, the pollen united to a pollinium stick to the tongue or the palpae of the moth, and the next flower to be visited by the insect will thus be pollinated. There is a suggestion that the length of the proboscis and the development of the spur have been acquired through co-adaptive evolution.

Since many tropical orchids have very long nectaries (in the well-known case of *Angraecum sesquipedale* from Madagascar it is more than 30 cm long), they have got into a real state of dependency on the animals pollinating them. The orchid just mentioned caused DARWIN in 1862 (cf. WARNER 1934) and WALLACE in 1891 (according to ROTHSCHILD & JORDAN 1903) to predict a moth with a particularly long tongue. This moth was later found and given the name *Xantho-pan morgani praedicta* by ROTHSCHILD & JORDAN (1903).

Coloration and markings have developed under the influence of different, part-ly antagonistic selective influences which are frequently countered by way of a clearly visible compromise.

In almost all sphingid species under consideration, the conspicuous coloration and markings are limited to the hindwings and sometimes only their basal parts, or they are restricted to lateral parts of the abdomen, such as in the case of *Manduca* or *Cocytius* (fig. 3), whereas the forewings and the thorax have a cryptic color-ation. During daytime, when the moth is in a resting position and could easily be seen, the forewings cover the conspicuously coloured hindwings and lateral ab-dominal parts completely.

In the typical resting position, the wings of the sphingids look roof-shaped, and the antennae are put along the body. In this position, the patterns of the fore-wings frequently form a perfect unit with those of the thorax or the abdomen (fig. 6b).

Among the Smerinthini there are species whose hindwings partly show up from underneath the front of the forewings while the moths are resting (fig. 5c). If disturbed, they move their forewings with a jerk (KERNBACH 1960), and the eye spots on the hindwings appear, which has a deterrent effect on possible enemies.

During daytime, the imagines are normally protected by their cryptic color-

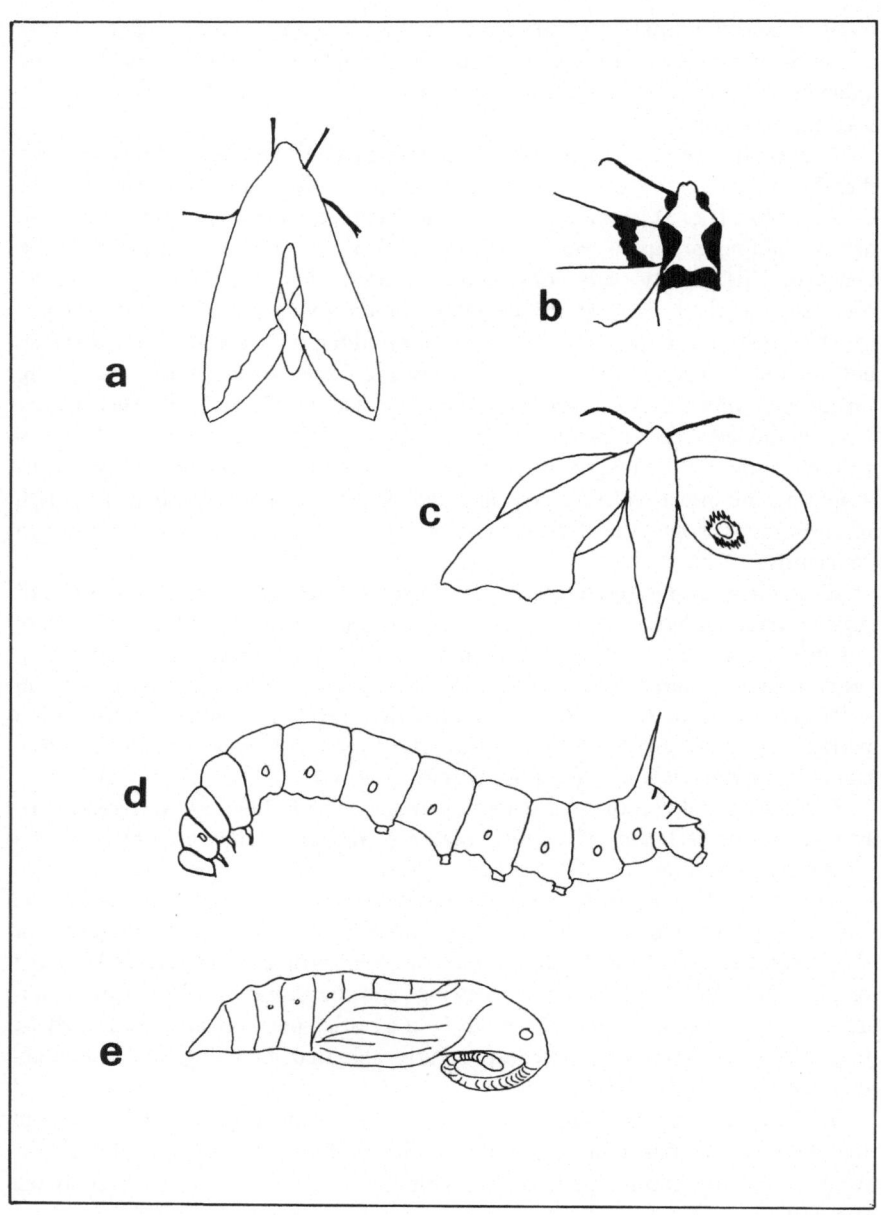

Fig. 6. Imago, larva and pupa: a) typical resting position of a *Sphinx*; b) markings complementing each other in resting position; c) typical resting position of a *Smerinthus*; d) caterpillar (from CARCASSON 1968); e) pupa.

ation and during the night, when most of the species are active, by their quickness, even though it happens that they are preyed upon by bats, geckos, giant toads and small mammals especially at light sources (SEITZ 1931). The smooth larvae, on the other hand, are in a particularly great need of protection from predators.

For this reason, the caterpillars have developed many different types of camouflaging and warning coloration (CURIO 1965). Often the first larval instars look completely different from the last. In some cases, the larvae show mimicry with snakes serving as models, such as the larvae of *Eumorpha labruscae, Madoryx pluto, Xylophanes tersa* (fig. 4.4), *Hemeroplanes ornatus* and *H. triptolemus*.

On the eleventh of a total of 13 segments, the caterpillars have a caudal horn, to which the English common name for sphingid caterpillars, 'hornworms', refers. In later larval instars, the caudal horn may be reduced; it is especially reduced to a tubercle in those cases, where the larvae imitate snakes. In *Lapara*, however, it is completely absent (HODGES 1971). In contradistinction to this, the caterpillar of *Ceratomia amyntor* has four horns on the thoracic rings.

According to FORSTER (1954), the caudal horn is a purely cutaneous structure resulting from two particularly strong bristles which have grown together. Its function is not clearly understood yet (CARCASSON 1968a). In view of the fact, however, that the horn has a bifid tip in the early instars, CARCASSON tends to believe that its function is sensory and that the horn might possibly help to deflect attack from the head in a similar way as do the tails of the hindwings or the eyespots of butterflies.

With the conspicuously, but aposomaticly coloured *Pseudosphinx tetrio* as well as some *Isognathus* species (fig. 4.5), the caudal horn is very long and thin. According to JORDAN (1911) it can be moved arbitrarily, and it could well be that it serves to get rid of parasites such as Jchneumonidae who like to lay their eggs in the larva (fig. 4.6.).

According to SEITZ (1927), some sphingid caterpillars are said to be protected by the poisonous substances contained in the plants on which they feed. Euphorbiaceae, Solanaceae and Apocynaceae are among the favourite foodplants of sphingids. KERNBACH (1957) wrote about the nauseating taste of the caterpillar of *Hyles euphorbiae*. In view of the fact, however, that even generally accepted examples of butterflies which have become inedible for their predators by feeding on poisonous plants have recently been contested, such as in the case of the Monarch (Danaidae) (URQUHART 1960), one should rather refrain from generalizing single observations made on sphingids.

In view of the length of the proboscis, the pupae frequently have a free tongue-case which, moreover, can be coiled up to two-and-one-half times such as in the case of *Neococytius cluentius* (HODGES 1971), whereas it is missing in *Pseudosphinx tetrio* and species of *Smerinthus* and *Pachysphinx* which are lacking a proboscis.

Pupation occurs in or on the ground; some South American species (JORDAN 1911) pupate under the bark of their foodtree.

Table 3. List of some North and South American sphingid species whose larval stages are not known or have not been described so far:

Cocytius beelzebuth		Erinnyis guttularis
Cocytius lucifer		Pachylia darceta
Manduca dilucida		Kloneus babayaga
Manduca occulta	(pupa unknown)	Madoryx pseudothyreus
Manduca stuarti		Callionima nomius
Manduca brontes	(pupa unknown)	Callionima pan
Dolbogene hartwegii		Callionima parce
Ceratomia sonorensis		Stolidoptera tachasara
Ceratomia hageni	(pupa unknown)	Pachygonia subhamata
Sagenosoma elsa	(pupa unknown)	Himantoides undata
Sphinx merops		Perigonia stulta
Sphinx lugens		Aellopos tantalus
Sphinx geminus		Aellopos clavipes
Sphinx eremitoides	(pupa unknown)	Hemaris gracilis
Sphinx istar		Hemaris senta
Sphinx chisoya		Eumorpha triangulum
Sphinx leucophaeata		Proserpinus clarkiae
Sphinx perelegans		Euproserpinus euterpe
Sphinx asella		Euproserpinus wiesti
Amplypterus palmeri		Xylophanes irrorata
Erinnyis domingonis		

The eggs are laid singly or in pairs on the underside of the leaves of the same plant (JORDAN 1911). CARY (1951) gives a detailed description of the life history of *Manduca sexta*.

Important bionomic data on many South American and also some North American species are still lacking. The foodplants of the caterpillars are partly still unknown, and there are species whose larval stages are not yet known or described (table 3) (cf. DRAUDT 1931, HODGES 1971, KERNBACH 1965, MOSS 1920, TIETZ 1972).

Furthermore, information about the early stages of species of the following genera seems to be lacking: *Nannoparce, Trogolegnum, Monarda, Protaleuron, Nyceryx, Phanoxyla*. Finally, there are species which even as imagenes are known only from few specimens, and accordingly there is a lack of information about early instars and the limits of their ranges.

In view of the particular correlation between sphingid species with long tongues and the flowers visited by them, it would be just as interesting to record the specific flowers which the imagines visit.

b. Classification

In their revision, ROTHSCHILD & JORDAN (1903) used the presence or lack of a basal scent patch at the inner side of the palpae as the decisive criterion for splitting up the Sphingidae into the Asemanophorae with the sub-families Acherontiinae and Ambulicinae and into the Semanophorae with the sub-families Sesiinae, Philampelinae and Choerocampinae.

DRAUDT in SEITZ (1931) followed this division, as did the SEITZ-work in general, in whose Palaearctic volume JORDAN dealt with the Sphingidae.

In recent works, too (DIERL 1970), this system has been maintained.

In contrast to this, CARCASSON (1968b) considered to recognize the categories Asemanophorae and Semanophorae as sub-families, HODGES (1971) pointed out that, according to the rules of the Code of Zoological Nomenclature, supergeneric units within a family must be based on a valid or an available generic name, which means to fulfil the demand for tautonomy between the name-giving genus and the higher categories of the family.

Table 4 shows the classification as used by HODGES (1971). Table 5 lists all sphingid genera of the New World together with the respective species representing the type of the genus.

A list of the species and subspecies can be found in the tabular survey of the distribution of the North and South American sphingids.

In the following, alterations as they result from a comparison of the quoted literature will be discussed in the sequence of the genera.

In his systematic revision of the Nearctic Sphingidae, HODGES (1971) synonymized the genus *Enyo* with *Epistor* and placed it near *Hemaris* on the grounds that he saw a close connection on the basis of features of the female genetalia and also in view of a similar habitus of the larvae. *Cautethia* has been expelled by HODGES (1971) from the Dilophonotini and relegated to the Macroglossini on the basis of features of the genitalia as well as the general appearance of the moth.

Himantoides, too, has been associated with the Macroglossini, because this genus shows close relations with *Cautethia* (DRAUDT 1931, p. 875). '*Cautethia* differs from the otherwise similarly built *Himantoides* (only) in having shorter

Table 4. Classification

			Genera	Name-giving genus	
Superfamily	Sphingoidea				
Family	Sphingidae	LATREILLE		Sphinx	LINNAEUS, 1758
1. Subfamily	Sphinginae	LATREILLE		Sphinx	LINNAEUS, 1758
Tribe 1	Sphingini	LATREILLE	1-16*	Sphinx	LINNAEUS, 1758
Tribe 2	Smerinthini	HÜBNER	17-25	Smerinthus	LATREILLE, 1802
2. Subfamily	Macroglossinae	HARRIS		Macroglossum	SCOPOLI, 1777
Tribe 1	Dilophonotini	BURMEISTER	26-46 **	Dilophonota	BURMEISTER, 1856
Tribe 2	Philampelini	BURMEISTER	47 ***	Philampelus	HARRIS, 1839
Tribe 3	Macroglossini	HARRIS	48-60	Macroglossum	SCOPOLI, 1777

* The numbering only reflects the sequence of the North and South American genera
** Synonym of Erinnyis HÜBNER
*** Synonym of Eumorpha HÜBNER

Table 5. Sphingid genera of the New World with type species

1. Agrius HBN., 1819	cingulatus F., 1775 (Sphinx)
2. Cocytius HBN., 1819	antaeus DRC., 1773 (Sphinx)
3. Neococytius HODGES, 1971	cluentius CRAM., 1776 (Sphinx)
4. Amphimoeca R. & J., 1903	walkeri BDV., 1875 (Amphonyx)
5. Manduca HBN., 1807	sexta JOH., 1763 (Sphinx)
6. Euryglottis BDV., 1875	aper WLK., 1856 (Macrosila)
7. Dolba WLK., 1856	hylaeus DRURY, 1773 (Sphinx)
8. Dolbogene R. & J., 1903	hartwegii BUTL., 1875 (Dolba)
9. Ceratomia HARR., 1839	amyntor GEY., 1835 (Agrius)
10. Isoparce R. & J., 1903	cupressi BDV., 1875 (Sphinx)
11. Nannoparce R. & J., 1903	poeyi GRT., 1865 (Hyloicus)
12. Sagenosoma JORD., 1946	elsa STKR., 1877 (Sphinx)
13. Neogene R. & J., 1903	reevi DRC., 1882 (Hyloicus)
14. Paratrea GRT., 1903	plebeja F., 1777 (Sphinx)
15. Sphinx L., 1758	ligustri L., 1758 (Sphinx)
16. Lapara WLK., 1856	bombycoides WLK., 1856 (Lapara)
17. Protambulyx R. & J., 1903	strigilis L., 1771 (Sphinx)
18. Amplypterus HBN., 1822	gannascus STOLL, 1790 (Sphinx)
19. Orecta R. & J., 1903	lycidas BDV., 1875 (Ambulyx)
20. Trogolegnum R. & J., 1903	pseudambulyx BDV., 1875 (Smerinthus)
21. Smerinthus LATR., 1802	ocellata L., 1758 (Sphinx)
22. Paonias HBN., 1819	excaecatus SMITH, 1797 (Sphinx)
23. Pachysphinx R. & J., 1903	modesta HARR., 1839 (Smerinthus)
24. Monarda DRC., 1896	oryx DRC., 1896 (Monarda)
25. Cressonia G&R, 1865	juglandis SMITH, 1797 (Sphinx)
26. Pseudosphinx BURM., 1856	tetrio L., 1771 (Sphinx)
27. Isognathus FLDR., 1862	caricae L., 1764 (Sphinx)
28. Erinnyis HBN., 1819	ello L., 1758 (Sphinx)
29. Phryxus HBN., 1819	caicus CRAM., 1777 (Sphinx)
30. Pachylia WLK., 1856	ficus L., 1758 (Sphinx)
31. Pachylioides HODGES, 1971	resumens WLK., 1856 (Pachylia)
32. Kloneus SKINNER, 1923	babayaga SKINNER, 1923 (Kloneus)
33. Oryba WLK., 1856	robusta WLK., 1856 (Oryba)
34. Hemeroplanes HBN., 1819	triptolemus CRAM., 1779 (Sphinx)
35. Madoryx BDV., 1875	oiclus CRAM., 1780 (Sphinx)
36. Callionima LUC., 1856	parce F., 1775 (Sphinx)
37. Stolidoptera R. & J., 1903	tachasara DRC., 1888 (Aleuron)
38. Protaleuron R. & J., 1903	rhodogaster R. & J., 1903 (Protaleuron)
39. Aleuron BDV., 1870	chloroptera PERTY, 1834 (Sphinx)
40. Pachygonia FLDR., 1874	subhamata WLK., 1856 (Perigonia)
41. Nyceryx BDV., 1875	hyposticta FLDR., 1874 (Ambulyx)
42. Perigonia H.-S., 1854	stulta H.-S., 1854 (Perigonia)
43. Eupyrrhoglossum GRT., 1865	sagra POEY, 1832 (Macroglossum)
44. Aellopos HBN., 1819	titan CRAM., 1777 (Sphinx)
45. Enyo HBN., 1819	lugubris L., 1771 (Sphinx)
46. Hemaris DALMAN, 1816	fuciformis L., 1758 (Sphinx)
47. Eumorpha HBN., 1807	labruscae L., 1758 (Sphinx)
48. Himantoides BUTL., 1877	undata WLK., 1856 (Perigonia)
49. Cautethia GRT., 1865	noctuiformis WLK., 1856 (Oenosander)
50. Gurelca KBY., 1880	hyas WLK., 1856 (Lophura)
51. Sphecodina BLCH., 1840	abbottii SWAINS., 1821 (Thyreus)

28

Table 5. (continued)

52. Deidamia CLEM., 1859	inscriptum HARR., 1839 (Pterogon)
53. Arctonotus BDV., 1852	lucidus BDV., 1852 (Arctonotus)
54. Amphion HBN., 1819	nessus CRAM., 1777 (Sphinx)
55. Proserpinus HBN., 1819	proserpina PALLAS, 1772 (Sphinx)
56. Euproserpinus G&R, 1865	phaeton G&R, 1865 (Euproserpinus)
57. Darapsa WLK., 1856	pholus CRAM., 1776 (Sphinx)
58. Xylophanes HBN., 1819	anubus CRAM., 1777 (Sphinx)
59. Phanoxyla R. & J., 1903	hystrix FLDR., 1774 (Chaerocampa)
60. Hyles HBN., 1819	gallii ROTTEMBURG, 1775 (Sphinx)

antennae and a normal scale pattern of the hind-tibiae'.* *Himantoides* is also very similar to *Pachygonia*, it is true, the latter, however (DRAUDT 1931), is 'also closely related with *Perigonia* and *Nyceryx*'.**

Darapsa has been replaced by HODGES (1971) within the Macroglossini; however, he expelled this genus from the Nephelicae (previously associated with Philampelinae) and, hence, from the tribe Philampelini, in which only *Eumorpha* and *Tinostoma**** remained, and placed it among the Macroglossini together with the remaining Nephelicae genera. In view of the extremely similar genitalia (HODGES 1971), *Darapsa* has been placed right before *Xylophanes*. The genus thus belongs to the category previously recognized as subfamily Choerocampinae (ROTHSCHILD & JORDAN 1903, DRAUDT 1931), which today would have to be named as subtribe Choerocampicae.

c. Nomenclature of sphingid genera of the New World

The North and South American sphingids are presently represented by 60 genera. This figure includes the alterations as suggested by HODGES (1971); as against the number of genera occurring in America as given by DRAUDT (1931) the number was increased by 2 through splitting — *Neococytius* partim *Cocytius* and *Pachilioides* partim *Pachylia* — and decreased by 4 through synonymizing *Chlaenogramma* with *Manduca*, *Isogramma* (= *Autogramma* JORDAN 1946) with *Ceratomia*, *Epistor* with *Enyo* and *Ampeloeca* with *Darapsa*.

In addition, according to the International Rules of Nomenclature, several generic names have changed as compared with the classification given by DRAUDT (1931).

The demand for necessary alterations of the nomenclature has been raised as

---

* *'Cautethia* unterscheidet sich von der sonst gleich gebauten *Himantoides* (lediglich) durch kürzere Fühler und normal beschuppte Hinterschienen'.
** 'auch mit *Perigonia* und *Nyceryx* nahe verwandt'.
*** from Hawaï, so far only known from few specimens.

early as in 1922 by BARNES & LINDSAY and also in 1946 by OITICICA; yet it has so far been largely ignored, as seen from the papers published since then. These alterations partly involve very well established names like *Herse, Protoparce, Pholus* and *Celerio* which either had to make way to the older names which have priority, or became invalid like *Celerio* and *Herse*, because the publication by OKEN (1815), in which these genus names have first been used, is not recognized. There are also other reasons responsible for alterations, like in the case of *Dictyosoma*, in which the name had originally already been used twice for other animal genera, before it was used by ROTHSCHILD & JORDAN (1903) to name a sphingid genus.

The following chapter will deal with the 13 sphingid genera of the New World, for which two or more names are being used, and discuss the reasons indicating which genus name is correct.

Agrius/Herse

Herse OKEN, 1815, appears for the first time in 'OKEN's Lehrbuch der Naturgeschichte 2 (Fleischlose Thiere)' 1815:762. After this work had been rejected in 1956 by a decision of the International Commission on Zoological Nomenclature in Opinion 417 on the grounds that it did not follow binary nomenclature, the next available name is *Agrius* HÜBNER [1819]* from the 'Verzeichniss bekannter Schmettlinge' [sic], p. 140, Augsburg.

Valid: *Agrius* HÜBNER [1819]

Manduca/Phlegethontius/Protoparce

This is a question of priority. *Manduca* HÜBNER [1807] is the oldest name. It appears in the 'Sammlung Exotischer Schmetterlinge' on plate [170]**, which corresponds with the first reference by HÜBNER in the 'Tentamen' (1806). *Phlegethontius* HÜBNER [1819] appears for the first time in 'Verzeichniss bekannter Schmettlinge' [sic], p. 140. HÜBNER mentions *Manduca* as synonym, as does the facsimile-edition of the HÜBNER-work 'Sammlung Exotischer Schmetterlinge' published by WYTSMAN [1904-1908] in Brussels. This work also states *Protoparce* as synonym, a name that goes back to BURMEISTER and appears in 1856 in 'Systematische Übersicht der Sphingidae Brasiliens, Sitzungsberichte', p. 63. Confusion was brought about by the fact that for a long time the exact dates of the HÜBNER publications have been unknown. This might well have been the reason why ROTHSCHILD & JORDAN (1903) considered *Manduca* as senior synonym of *Protoparce*. *Phlegethontius* was rejected by them on the grounds that HÜBNER included *cluentius* CRAMER as type of *Cocytius*.

The work of FRANCIS HEMMING (1937): 'HÜBNER: A bibliographical and

---

* The work was published in parts between 1816 and 1826; the sphingids are dealt with in the part published in 1819.
** The square bracket means here that the plates which have been sold in several parts (1806-1838), have originally not been numbered consecutively by HÜBNER.

systematic account of the entomological works of JACOB HÜBNER ...' fixed the dates of publication of the different plates of the 'Sammlung exotischer Schmetterlinge' and specified the year 1807 for *Manduca*.

Even though Opinion 97* says that the generic names specified in the 'Tentamen' are not binding, this does not apply to 'Sammlung Exotischer Schmetterlinge' as well. The demand for sufficient description is fulfilled by the figure.

The objection that the well-known name *Protoparce* should be maintained for reasons of stability has to be rejected on the very ground that three different generic names are presently in use. A decision thus necessary should be made in favour of the oldest available name.

Valid: *Manduca* HÜBNER [1807]

Sagenosoma/Dictyosoma

The genus *Dictyosoma* has been established by ROTHSCHILD & JORDAN (1903). OITICICA (1946) pointed out in 'Revisão Dos Nomes Genéricos Da Família Sphing. (Lep.)', Boletim Do Museu Nacional, Rio de Janeiro, Zoologia N. 66, that this is an invalid junior homonym which he replaces by the name *Sagenosoma* designated by JORDAN (1946). *Dictyosoma* is preoccupied by Prot. 1856 and Pisces 1877.

Valid: *Sagenosoma* JORDAN, 1946

Paratrea/Atreides/Atreus

Both GROTE and HOLLAND mentioned (1903) that *Atreus* GROTE, 1886, is an invalid homonym of *Atreus* KOCH, 1837, Arachnida. GROTE suggests the new name *Paratrea* published in 'Canad. Entomology', XXXV:207, No. 7, July 1903. HOLLAND's suggestion to replace *Atreus* by *Atreides* is found in 'The Moth Book', p. 49. The preface by HOLLAND bears the date of 8th September, 1903. Thus, *Paratrea* is older than *Atreides*.

Valid: *Paratrea* GROTE, VII 1903

Paonias/Calasymbolus

Paonias HÜBNER [1819] is to be found in the 'Verzeichniss bekannter Schmettlinge' [sic]:142 and includes the *Smerinthus* species *s(alicis) ocellata* L. mentioned in the first place. ROTHSCHILD & JORDAN (1903) list this species under *Sphinx* L. Since these authors advocated the 'first species principle', they understood *Smerinthus s. ocellata* L. as being designated as type of *Paonias* in the quoted passage and mentioned *Paonias* as junior synonym of *Sphinx*, whereas they used *Calasymbolus* GROTE, 1877 (1873 nom. indescr.) for *Paonias* HÜBNER [1819] partim.

The principle applied by ROTHSCHILD & JORDAN is, however, not sup-

---

* Decision of the International Commission on Zoological Nomenclature, published in Science 1927, LXV:301.

ported by a corresponding rule of the Code*. In splitting a taxon, a valid name has to pass to one of the components.

Valid: *Paonias* HÜBNER [1819]

Phryxus/Grammodia
In the 'Verzeichniss bekannter Schmettlinge' [sic] *Phryxus* HÜBNER [1819] is likewise a heterogeneous genus. In splitting it up ROTHSCHILD & JORDAN (1903) argue like in the preceding case relegating *Phryxus* with the species *livornica* as synonym to *Celerio* and introducing *Grammodia* R. & J. 1903 as nomen novum for the remaining species. *Grammodia* is therefore a junior synonym of *Phryxus* (cf. BARNES & LINDSAY (1922).

Valid: *Phryxus* HÜBNER [1819]

Kloneus/Oberthurion
It seems reasonable in this case to consider which of the two species names of this monotypic genus is the only correct name of this species which two authors, SKINNER & CLARK, established independent of each other in 1923 almost at the same time.

According to LICHY (1945) (cf. also KERNBACH 1965a) SKINNER is said to have described the species under the name of *Kloneus babayaga* on May 4th, 1923. The publication is contained in the 'Entomological News and Proceedings of the Entomological Section of the Academy of Natural Sciences of Philadelphia', 1923, 34:138. In contrast to the statement by KERNBACH (1965a), No. 5 bears the date '23rd. May 1923' according to my examination. CLARK described and published the species on May 12th, 1923 in 'Proceedings of the New England Zoological Club', Cambridge, 1923, Vol. VIII, p. 58, as *Oberthurion harroverii*. In accordance with this, this name is also used by DRAUDT (1931).

Although (cf. III:Article 9(5) of the 'International Code of Zoological Nomenclature') 'labelling of a specimen in a collection' '... does not constitute publication', the labelling of the specimen which is now in Carnegie Museum, Pittsburgh, Pa., USA, is thought-provoking. The label in CLARK's handwriting bears the mention 'sunk' with regard to the name established by him.

According to the information obtained by KERNBACH (KERNBACH 1965a) from the Academy of Natural Sciences of Philadelphia, the latter had received the publication of SKINNER as early as on May 9th, 1923. The mailing date must therefore have been earlier, most probably on May 4th, 1923 (cf. LICHY 1945).

It says, however, in the interpretation of the rules of nomenclature**: 'The date of publication is the date on which the publication was mailed to subscribers, placed on sale, or, where the whole edition is distributed free of charge, mailed to institutions and individuals to whom such free copies are normally distributed.

---

* International Code of Zoological Nomenclature published by the Int. Trust for Zool. Nomencl. London 1961 (rev. ed. 1964).
** MAYR 1969a, p. 349.

Journals are sometimes mailed weeks or even months after the date printed on the covers. In these cases the mailing date is the correct date, not the printed date'.

In the case in question, the correspondingly decisive mailing date would be anterior to the printed date, which, even though confusing, is not relevant in this case.

Valid: *Kloneus babayaga* SKINNER, 4th (?) May, 1923

Hemeroplanes/Leucorhampha

Hemeroplanes HÜBNER [1819] from 'Verzeichniss bekannter Schmettlinge' [sic]:133 included species which have meanwhile been assigned to several genera. KIRBY (1892), in 'A Synonymic Catalogue of Lepid.Het.':647, unequivocally established *triptolemus* CRAMER marked with an asterisk – as type of *Hemeroplanes*, wich was confirmed by BARNES & LINDSAY (1922). The name *Leucorhampha* later introduced as nomen novum by ROTHSCHILD & JORDAN (1903) has, therefore, to be considered as junior synonym.

Valid: *Hemeroplanes* HÜBNER [1819]

Callionima/Calliomma/Hemeroplanes

In view of the fixation of *Hemeroplanes* HÜBNER [1819] through the type-designation by KIRBY, ROTHSCHILD & JORDAN (1903) wrongly used this name. Therefore, the next available name must be reverted to. *Calliomma* WALKER, 1856 (July 28th) appears in 'List of the Specimens of Lepidopterous Insects in the Collection of the British Museum' 8:108, quoted as a manuscript name of BOISDUVAL. The same reference is to be found with LUCAS (1857) in 'RAMON DE LA SAGRA, Histoire Physique, Politique et Naturelle de l'Ile de Cuba', p. 680, where it says *Callionima* and in a footnote BOISD. (gen.nov.ined.) A Spanish edition of this work appeared in 1856. *Callionima* is to be found on page 291. It is superfluous, however, to clarify the question of priority, because, as HODGES (1971) says, *Calliomma* WALKER, 1856, is a homonym of *Calliomma* AGASSIZ, 1846, an Arachnidae genus.

Valid: *Callionima* LUCAS, 1856

Aellopos/Sesia

Sesia has been established in 1775 by FABRICIUS in 'Systema Entomologiae ...':547, where it comprises five species of Sphingidae and four species of the family Sesiidae. There are several attempts documented in literature either to replace, within the sphingids, *Sesia* by *Aellopos* HÜBNER [1819]:131, 'Verzeichniss bekannter Schmettlinge' [sic], or to rename the family Sesiidae together with the tautonymous type genus into Aegeriidae and *Aegeria* FABRICIUS, 1897, respectively, which points to the difficulty to find a reference to the type-designation. Starting from the year 1964, the Zoological Record refers to Sesiidae as a family name again.

The homonymous use of *Sesia* FABRICIUS for a sphingid genus has to be prevented through replacement by the next available name.

Valid: *Aellopos* HÜBNER [1819]

Hemaris/Haemorrhagia
Hemaris DALMAN, 1816, in 'Kongl. Vetenskaps Academiens Handlingar' XXXVII: 207 is older than *Haemorrhagia* GROTE & ROBINSON, 1865, from 'A Synonymical Catalogue of North American Sphingidae ...' in: Proc. Ent. Soc. Philad. 5:173. ROTHSCHILD & JORDAN (1903) specify *Hemaris* as synonym on the grounds that *Hemaris* included the type of *Macroglossum*. BARNES & LINDSAY (1922) point to the fact that, according to Opinion 10 of the 'International Commission on Zoological Nomenclature', the applicability of *Hemaris* can be inferred from analogous instances.

Valid: *Hemaris* DALMAN, 1816

Eumorpha/Pholus
Eumorpha HÜBNER [1807] in 'Sammlung exotischer Schmetterlinge', plate [167]* has priority over *Pholus* HÜBNER [1819], 'Verzeichniss bekannter Schmettlinge' [sic]:134. What has been said in connection with *Manduca* also applies to this case.

Valid: *Eumorpha* HÜBNER [1807]

Hyles/Spectrum/Celerio
Celerio OKEN, 1815, is invalid on the grounds that the work 'OKEN's Lehrbuch der Naturgeschichte ...' has been rejected (Opinion 417, 1956, Int. Comm. on Zool. Nomenclature). The next available name is *Hyles* HÜBNER [1819], from 'Verzeichniss bekannter Schmettlinge' [sic]:137. FLETCHER (1966) published *Spectrum* SCOPOLI, 1777, as the genus name that replaces *Celerio*. In SCOPOLI's work 'Introductio ad Historiam Naturalem ...':413, *Spectrum* contains species which are now split up among several genera and included the type of *Sphinx*, *Sphinx ligustri* LINNAEUS.

OITICICA (1946) had already rejected the applicability of *Spectrum* on the grounds that it is the 'isogenotipico' (OITICICA) of *Sphinx* LINNAEUS, 1758, and hence an unavailable genus name.

Valid: *Hyles* HÜBNER [1819]

---

* Plate numbering not by HÜBNER

34

V. CHOROLOGY

a. The general distribution of the Sphingidae and the distribution of the American genera

The Sphingidae have a world-wide distribution as shown in BARTHOLOMEW et al. (1911), the range of the family leaving open only the extreme north and south of the earth. This points to a certain climatic dependency, although this is no criterion by which the Sphingidae would, in terms of their distribution, differ greatly from many other animal groups with similarly wide ranges. Comparing, however, the number of species that occur on the northern and southern continents respectively, it becomes evident that the Sphingidae have a predominantly tropical distribution. Nevertheless, they have been able to settle on island groups far from the continent, such as the Bermudas, the Galápagos Islands and Hawaii, unless low temperatures had a hampering effect on their spreading.

The family of Sphingidae comprises approximately 1000 species on a world-wide level (HODGES 1971). Only 21 species occur in Central Europe, of which as many as 8 species are strays or regular migrants from the south.

The number of species occurring in the whole of Europe amounts to 28, as against 260 species occurring in Africa (CARCASSON 1968a). In the New World the Sphingidae are represented by 60 genera including 432 species.

As shown in table 6, 20 of these genera are restricted to the Neotropical region. In comparison, only 10 genera occur exclusively in the Nearctic region. 7 further genera distinguish themselves by the vast ranges of some of their species. *Xylophanes tersa* e.g. occurs from Canada to Argentina. In the ratio of 13:8, 21 genera contain species which, either coming from the Neotropical region just reach the southern border of the Nearctic region, especially southern Florida (cf. *Phryxus caicus*, fig. 30, p. 138, and *Pachylia ficus*, fig. 31, p. 139) or, coming from North America are still found south of the faunal border-line.

HODGES (1971) mentions 40 genera for the Nearctic region; this figure, however, also includes genera, some of whose species occur only as strays in North America (cf. *Neococytius cluentis*, fig. 32, p. 140, *Pachylioides resumens*, fig. 33, p. 141).

It might, at first glance, seem somewhat inconsistent to incorporate *Dolbogene, Ceratomia, Proserpinus* and *Darapsa* in column 4. Yet, I consider this justified, since in the case of species with small ranges, a single locality is of greater importance, especially if, as in the case of *Proserpinus vega mooseri*, only very few specimens have so far been recorded (cf. fig. 34, p. 142). *Dolba* has been placed among the Nearctic genera, even though, besides the North American *Dolba hyloeus* (DRURY), *Dolba schausi* has been described by CLARK (1916) on the basis of a singleton from the Cauca Valley in Colombia (type ♀, in CM, Pittsburgh). According to HODGES (1971), *Dolba* is a monotypic genus.

Euproserpinus is likewise listed as a Nearctic genus, although, in the British Museum, there is one specimen of *Euproserpinus euterpe* from Guerrero, Mexico, and another from Solola (Guatemala?), which I registered as *Euproserpinus spec.* only. *Smerinthus* and *Pachysphinx* are actually Nearctic genera; species reaching Mexico are confined essentially to the Plateau. This is also true of the species observed by CARY (1963) in Baja California. The author writes about *Smerinthus cerisy* [sic] *opthalmatica* [sic] BDV. sensu CARY: 'It is most interesting to find that this southern California subspecies extends its range along the peninsula and that it appears to be commoner in the mountains'. In view of the fact that *Smerinthus saliceti* has been observed in Guadalajara/Jalisco, Orizaba and Hidalgo, and *Pachysphinx modesta regalis* has likewise been found in Tehuacan and Guadalajara/Jalisco, *Smerinthus* and *Pachysphinx* have been included in column 4. The symbols in brackets indicate that *Agrius, Sphinx, Smerinthus, Hemaris, Gurelca, Sphecodina, Proserpinus* and *Hyles* are genera which include species occurring also in the Old World. *Hyles galii* occurs in the Palaearctic and in the Nearctic regions, in the latter as form *intermedia* according to HODGES (1971). *Sphinx pinastri* has been brought in from Europe to North America, *Hyles euphorbiae* from Europe to Canada (HODGES 1971).

Table 6. Proportional distribution of sphingid genera in the New World 1. exclusively Neotropical genera; 2. Nearctic genera; 3. genera with far ranging species sharing both faunal regions of the New World; the arrows indicate the direction of dispersal, which is, however, not necessarily that of the latest expansive phase; 4. genera, some of whose species share the adjacent faunal region marginally (symbols in brackets indicate that the respective genera are also represented in the Old World).

Genus	1	2	3	4
1 Agrius			(↑)	
2 Cocytius				↑
3 Neococytius	X			
4 Amphimoeca	X			
5 Manduca			↑	
6 Euryglottis	X			
7 Dolba		X		
8 Dolbogene				↑
9 Ceratomia				↓
10 Isoparce		X		
11 Nannoparce	X			
12 Sagenosoma		X		
13 Neogene	X			

36

Table 6. (continued)

Genus	1	2	3	4
14 Paratrea				↓
15 Sphinx			(↓)	
16 Lapara		X		
17 Protambulyx				↑
18 Amplypterus	X			
19 Orecta	X			
20 Trogolegnum				↑
21 Smerinthus				(↓)
22 Paonias		X		
23 Pachysphinx				↓
24 Monarda		X		
25 Cressonia		X		
26 Pseudosphinx				↑
27 Isognathus	X			
28 Erinnyis			↑	
29 Phryxus				↑
30 Pachylia				↑
31 Pachylioides	X			
32 Kloneus	X			
33 Oryba	X			
34 Hemeroplanes	X			
35 Madoryx				↑
36 Callionima				↑
37 Stolidoptera	X			
38 Protaleuron	X			
39 Aleuron	X			
40 Pachygonia	X			
41 Nyceryx	X			
42 Perigonia				↑
43 Eupyrrhoglossum	X			
44 Aellopos				↑
45 Enyo				↑
46 Hemaris		(X)		
47 Eumorpha			↑	
48 Himantoides	X			
49 Cautethia				↑
50 Gurelca				(↓)
51 Sphecodina		(X)		
52 Deidamia		X		
53 Arctonotus		X		
54 Amphion				↓
55 Proserpinus				(↓)
56 Euproserpinus		X		
57 Darapsa				↓
58 Xylophanes			↑	
59 Phanoxyla	X			
60 Hyles			(↓)	

b. Tabular survey of the distribution of North and South American Sphingidae

The following tabular survey illustrates the distribution of North and South American sphingid species and subspecies, as it has been compiled according to the localities recorded from museum specimens and supplemented by distribution data obtained from literature.

Such a tabular survey can, of course, give but an approximate pattern of distribution, its unequivocal advantage, however, lies in the compact illustration of the distribution patterns which enables a survey to be gained of all the species and subspecies (as understood in the present work) which occur in the area under investigation as well as the adjacent Nearctic region. Moreover, with this illustration, one can quickly tell apart taxa with a wide distribution range from those with a narrow one. Finally, this survey offers a first possibility of comparison with regard to the centres of distribution to be worked out.

Interrupted distribution patterns may result from technical reasons of presentation and must not be taken for examples of disjunct ranges. Moreover, from markings of distribution one beneath the other one cannot straight away infer that the two taxa have a sympatric occurrence in the area in question (which would reduce any subspecies within one and the same species to absurdity), because the geographical regions specified in the different columns cover large areas for the most part, let alone their ecological differentiation.

Table 7. Systematic-chorological survey of the Sphingidae occurring in North and South America. The occurrence of a species or subspecies in a geographical region has been marked by a cross (+) and, in cases of vast ranges, by an uninterrupted line. A dot has been used to mark localities outside the main area of distribution as well as for taxa known only from one specimen. In some instances, the following abbreviations had to be used: A = Arizona, BA = Buenos Aires, BC = Baja California, CR = Costa Rica, E = East, F = Florida, G = Guatemala, GC = Grand Cayman I., H = Honduras, LC = Little Cayman I., M = Martinique, MW = Middle-West, N. = North, NE = North-East, NW = North-West, P. = Panama, RG = Revilla Gigedo I., SE = South-East, SI = Socorro I., SW = South-West, SJ = St. John I., T = Texas, W = West, Y = Yucatán.

Table 7. Distribution

Distribution matrix (columns correspond to the numbered species listed below):

Locality	1	2	3	4	5	6	7	8	9	10	11	12	13	14	15	16	17	18	19	20	21	22	23	24	25
Canada	•												+												+
North USA	•												+												+
West USA	+																								
East USA																									
South USA										F															
Mexico	•		+			+				F					+										
Central America	+		+										+												+
Guiana						+									+										
Venezuela															+								•		
Colombia				+																					
Ecuador				+					•																
Peru		+		•													+				+				•
Bolivia				+																					
North Brazil	+									+	+												•		
West Brazil				+																					
East Brazil	SE					SE					+														
South Brazil	+	+				+					+							•							
Paraguay																									
Uruguay																									
Argentina			+							+	+					+	+								
Chile										+							+		•						
Trinidad			+			+				+					+										
Cuba					+						•												+		•
Jamaica					•						+														
Hispaniola										+	•										+	+		+	
Puerto Rico																					+	+			
Lesser Antilles																									
Bahamas			+							+			+											+	
Galápagos	+																		+						

Species

1. Agrius cingulatus (F.)
2. Cocytius beelzebuth (BDV.)
3. duponchel (POEY)
4. mortuorum R.&J.
5. vitrinus R.&J.
6. lucifer lucifer R.&J.
7. lucifer lindneri GEHLEN
8. macasensis CLARK
9. antaeus antaeus (DRURY)
10. antaeus medor (STOLL)
11. Neococytius cluentius (CRAM.)
12. Amphimoeca walkeri (BDV.)
13. Manduca sexta sexta (L.)
14. sexta jamaicensis BUTL.
15. sexta paphus (CRAM.)
16. sexta peruviana BRYK
17. sexta saliensis KERNBACH
18. sexta caestri (BLCH.)
19. sexta leucoptera (R.&J.)
20. mossi (JORD.)
21. caribbea (CARY)
22. afflicta afflicta (GRT.)
23. afflicta bahamensis (CLARK)
24. afflicta johanni (CARY)
25. quinquemaculata (HAW.)

39

Table 7. (continued)

Distribution

	M. dilucida	azteca	kuschei	lucetius lucetius	lucetius nubila	lucetius exiguus	lucetius panaquire	reducta	diffissa diffissa	diffissa petuniae	diffissa tropicalis	diffissa ochracea	diffissa mesosa	diffissa zischkai	jordani	occulta occulta	occulta pacifica	empusa	hannibal hannibal	hannibal hamilcar	hannibal mayeri	pellenia pellenia	pellenia janeira	leucophila	perplexa
Canada																									
North USA																									
West USA																									
East USA																									
South USA																	•								
Mexico	+	+	ᴡ							•	+					+	ᴡ	+		+	+	+	•		
Central America	+			+							+					+						+			
Guiana		•																					+		
Venezuela	+	+								•						•		+							
Colombia																•					+	+			
Ecuador		•								•															
Peru		+			+	•	•	+								•									
Bolivia		+		+	•	•								+	•			+	+				+		
North Brazil									+														+		
West Brazil		•				•										+	+								
East Brazil		ꜱꜱ				ꜱꜱ	+			ꜱꜱ						•	ꜱꜱ								
South Brazil						•	+									•	+					ꜱꜱ		+	
Paraguay				+						•							+								
Uruguay			+			+																			
Argentina		+		+	+	+	ʙᴀ	ɴ	+	•						•								
Chile			+																						
Trinidad						+					+		+												
Cuba		•								•															
Jamaica																									
Hispaniola		•																							
Puerto Rico																									
Lesser Antilles																									
Bahamas																									
Galápagos																									

Species

Manduca dilucida (HY.EDW.)
azteca (MOOSER)
kuschei (CLARK)
lucetius lucetius (STOLL)
lucetius nubila (R.&J.)
lucetius exiguus (GEHLEN)
lucetius panaquire (BERG)
reducta (GEHLEN)
diffissa diffissa (BUTL.)
diffissa petuniae (BDV.)
diffissa tropicalis (R.&J.)
diffissa ochracea (CLARK)
diffissa mesosa (R.&J.)
diffissa zischkai (KERNBACH)
jordani (GIACOMELLI)
occulta occulta (R.&J.)
occulta pacifica (MOOSER)
empusa (KERNBACH)
hannibal hannibal (CRAM.)
hannibal hamilcar (BDV.)
hannibal mayeri (MOOSER)
pellenia pellenia (H.-S.)
pellenia janeira (JORD.)
leucophila (GEHLEN)
perplexa (R.&J.)

Table 7. (continued)

Distribution

Distribution matrix (locations × species). Symbols: + = present, • = record, | = range line, SE / BC = regional abbreviations.

Locality	scutata scutata (R.&J.)	scutata boliviana (CLARK)	scutata brasiliensis (JORD.)	clarki (R.&J.)	tucumana (R.&J.)	ochus (KLUG)	lefeburei lefeburei (GUÉR.)	lefeburei bossardi (GEHLEN)	incisa incisa (WLK.)	incisa pallidula (DANIEL)	incisa prestoni (GEHLEN)	andicola (R.&J.)	stuarti (ROTHSCH.)	manducoides (ROTHSCH.)	viola-alba (CLARK)	albolineata (GEHLEN)	brunalba (CLARK)	rustica rustica (F.)	rustica cortesi (CARY)	rustica harterti (ROTHSCH.)	rustica dubana (WOOD)	rustica dominicana (GEHLEN)	rustica calapagensis (HOLL.)	chinchilla (GEHLEN)	centrosplendens (GEHLEN)	albiplaga albiplaga (WLK.)
Canada																										
North USA																										
West USA																										
East USA																										
South USA																		+								
Mexico	+						+	•	+	•	•	+						|	BC							•
Central America	+						+	+										|								+
Guiana								•	•									|								
Venezuela	+						+		+									|								
Colombia	|			+			|		|									|								
Ecuador	|			|		+	|		|			+						|								
Peru	|			|					•								•	•						•	•	
Bolivia	•	+		+					•		•						+	+						•	•	
North Brazil									•									|								
West Brazil									•		+	+					+	+	•	•						+
East Brazil			SE						SE								+	+	•					SE		•
South Brazil	•	+							+		•	•				SE										SE
Paraguay		•							•			+					+	+		•						
Uruguay																										
Argentina						+										+	+									
Chile															+											
Trinidad					+	+												+								+
Cuba																					+					
Jamaica																	+									
Hispaniola																						+				
Puerto Rico																						+				
Lesser Antilles																				+						
Bahamas																										
Galápagos																							+			

Species: *Manduca*

Table 7. (continued)

Distribution

The table records geographic distribution (rows, left-hand column) for each species (columns). Symbols: + or • = present; letter codes (F, A, G, CR, SE) as printed.

Regions (top to bottom): Canada · North USA · West USA · East USA · South USA · Mexico · Central America · Guiana · Venezuela · Colombia · Ecuador · Peru · Bolivia · North Brazil · West Brazil · East Brazil · South Brazil · Paraguay · Uruguay · Argentina · Chile · Trinidad · Cuba · Jamaica · Hispaniola · Puerto Rico · Lesser Antilles · Bahamas · Galápagos

Species	Distribution (regions with marks)
Manduca albiplaga exacta (GEHLEN)	West Brazil
trimacula (R.&J.)	Colombia, Bolivia
leucospila (R.&J.)	Peru, Bolivia, North Brazil, West Brazil, East Brazil (SE), South Brazil (SE)
dalica dalica (KBY.)	Central America (CR), Venezuela, Bolivia
dalica anthina (JORD.)	Ecuador
composi (SCHAUS)	South USA (F), Cuba, Jamaica, Hispaniola, Puerto Rico, Bahamas
brontes brontes (DRURY)	Mexico, Central America (G)
brontes cubensis (GRT.)	Central America (CR)
brontes haitiensis (CLARK)	Mexico, Central America
brontes smythi (CLARK)	South USA (A)
sesquiplex sesquiplex (BDV.)	Colombia, Ecuador, Peru, Bolivia
sesquiplex opima (R.&J.)	Venezuela
muscosa (R.&J.)	Mexico, Central America, Venezuela
bergi (R.&J.)	Uruguay, Argentina
bergamatipes (CLARK)	Argentina
armatipes (R.&J.)	Uruguay, Argentina
carrerasi (GIACOMELLI)	Argentina
corallina (DRC.)	Mexico, Central America, Venezuela, Bolivia, South Brazil, Argentina
extrema (GEHLEN)	Mexico, Central America, Venezuela, Peru, Bolivia
lichenea (BURM.)	South USA (F), Central America, South Brazil, Argentina
feronia (KERNBACH)	South USA, Bolivia, West Brazil, South Brazil, Paraguay, Uruguay, Argentina, Trinidad, Cuba, Hispaniola
florestan florestan (CRAM.)	South USA, Mexico, Venezuela, Argentina
florestan vogli (DANIEL)	Mexico, Venezuela, Argentina
florestan ishkal (SCHAUS)	Venezuela
florestan argentinica (DANIEL)	Mexico, Argentina

Table 7. (continued)

Distribution

Region	grandis (DANIEL)	vestalis (JORD.)	fosteri (R.&J.)	lanuginosa (HY.EDW.)	crocala (DRC.)	barnesi (CLARK)	holcombi (MOOSER)	corumbensis (CLARK)	maricina (SCHAUS)	huascara (SCHAUS)	jasminearum (GUÉR.)	undata undata (R.&J.)	undata cinerea (R.&J.)	obscura (CLARK)	suavis HODGES	gueneei (CLARK)	Euryglottis albost. albostigmata ROTHSCH.	albostigmata basalis ROTHSCH.	dognini ROTHSCH.	davidianus DOGN.	aper (WLK.)	guttiventris R.&J.	Dolba hyloeus (DRURY)	Dolbogene hartwegii (BUTL.)	Ceratomia hageni GRT.
Canada													•											•	
North USA													+											+	
West USA																									MW
East USA													+										+		
South USA														•									+	+	+
Mexico					+	+	+	+															•	•	+
Central America		•			+	+	•							•											+
Guiana		•																							
Venezuela		+		+							+									+					
Colombia																	•					+			
Ecuador				+													•	•		+	+	+			
Peru																		+			+				
Bolivia																				+	+	+			
North Brazil	+																								
West Brazil									+																
East Brazil																									
South Brazil		•														+									
Paraguay		•																							
Uruguay																+									
Argentina		+													+	+									
Chile																									
Trinidad																									
Cuba																								•	
Jamaica																									
Hispaniola																									
Puerto Rico																									
Lesser Antilles																									
Bahamas																									
Galápagos																									

Species

Table 7. *(continued)*

Distribution

Species	*Ceratomia amyntor* (GEY.)	*undulosa* (WLK.)	*sonorensis* HODGES	*hoffmanni* MOOSER	*igualana* SCHAUS	*catalpae* (BDV.)	*Isoparce cupressi* (BDV.)	*Nannoparce poeyi poeyi* (G.&R.)	*poeyi haterius* (DRC.)	*balsa* SCHAUS	*Sagenosoma elsa* (STKR.)	*Neogene reevi reevi* (DRC.)	*reevi minor* CLARK	*pictus* CLARK	*intermedia* CLARK	*curitiba* JONES	*dynaeus dynaeus* (HBN.)	*dynaeus corumbensis* CLARK	*steinbachi* CLARK	*albescens* CLARK	*Paratrea plebeja* (F.)	*Sphinx arthuri* ROTHSCH.	*maura* BURM.	*aurigutta* (R.&J.)	*phalerata* KERNBACH
Canada	+	+																							
North USA	+	+				+																			
West USA																									
East USA	+	+				+	SE														+	+			
South USA		+	A			F	+			A											F		A		
Mexico				+	+		•	Y	+	N															
Central America																									
Guiana																									
Venezuela																									
Colombia																									
Ecuador																				•					
Peru																								+	
Bolivia																		+			+	+		+	
North Brazil																									
West Brazil															+		+								
East Brazil																	+			•					
South Brazil												+			SE					+					
Paraguay												+	+	+											
Uruguay																									
Argentina												+		•				•	+			+	•		+
Chile																								+	
Trinidad																									
Cuba								+																	
Jamaica								+																	
Hispaniola								+																	
Puerto Rico																									
Lesser Antilles																									
Bahamas																									
Galápagos																									

Table 7. (continued)

Distribution

Species	Sphinx justiciae WLK.	merops merops BDV.	merops monjena (SCHAUS)	merops judsoni (WEISS)	tricolor CLARK	lugens WLK.	geminus (R.&J.)	balsae SCHAUS	biolleyi (SCHAUS)	eremitus (HBN.)	pitzahuac MOOSER	pseudostigmatica GEHLEN	eremitoides STKR.	separata separata NEUM.	separata melaena R.&J.	istar (R.&J.)	xantus CARY	chisoya (SCHAUS)	praelonga (R.&J.)	leucophaeata CLEM.	chersis chersis (HBN.)	chersis mexicana R.&J.	adumbrata (DYAR)	vashti STKR.	libocedrus libocedrus HY.EDW.
Canada										+												+			+
North USA										+															
West USA																									+
East USA						•				+															
South USA						•	•							+	+	•	+		+		•	+		+	+
Mexico	+				+	+	+	+			+	+		+	+			BC	+		•	+		+	+
Central America	+				+	+		+					•			+					+	•			
Guiana																									
Venezuela		+													•										
Colombia		|																							
Ecuador		+	+	+																					
Peru		•																							
Bolivia		+																							
North Brazil																									
West Brazil																									
East Brazil		+																							
South Brazil	+																								
Paraguay																									
Uruguay																									
Argentina	+																								
Chile																									
Trinidad																									
Cuba		•																							
Jamaica																									
Hispaniola					+																				
Puerto Rico																									
Lesser Antilles					+																				
Bahamas																									
Galápagos																									

Table 7. *(continued)*

Species	Canada	North USA	West USA	East USA	South USA	Mexico	Central America	Guiana	Venezuela	Colombia	Ecuador	Peru	Bolivia	North Brazil	West Brazil	East Brazil	South Brazil	Paraguay	Uruguay	Argentina	Chile	Trinidad	Cuba	Jamaica	Hispaniola	Puerto Rico	Lesser Antilles	Bahamas	Galápagos
Sphinx libocedrus achotla MOOSER						+																							
perelegans HY.EDW.	+	+	+		+																								
asella (R.&J.)				+	+																								
canadensis BDV.	+	+		+																									
franckii NEUM.			MW	+																									
kalmiae J.E.SMITH	+	+		+																									
gordius CRAM.	+	—	—	+	F																								
luscitiosa CLEM.	+	—	—	+																									
drupiferarum J.E.SMITH	+	—	—	+																									
dollii NEUM.			+	—	+																								
sequoiae BDV.			+		+																								
pinastri pinastri L.	•	NW	•																										
Lapara coniferarum (J.E.SMITH)	+	+		+	F																								
bombycoides WLK.	+	+		+	F																								
Protambulyx eurycles (H.-D.)						+	—	—	—	—	—	—	—	—	—	—	—	+		+		+							
fasciatus GEHLEN																	•												
xanthus xanthus R.&J.							+																						
xanthus australis CLARK																	+												
euryalus R.&J.								•	+	—	—	—	+	•															
ockendeni R.&J.										+		+																	
sulphureus (ROTHSCH.)									+	+		•																	
astygonus (BDV.)						•		•	+			•	•	•			SE	+		+									
goeldii goeldii R.&J.							CR	+	+			+		+			•												
goeldii andicus GEHLEN										W																			
strigilis strigilis (L.)					•	+	—	—	—	—	—	—	—	—	—	—	—	—	—	+		+	—	—	—	—	+		

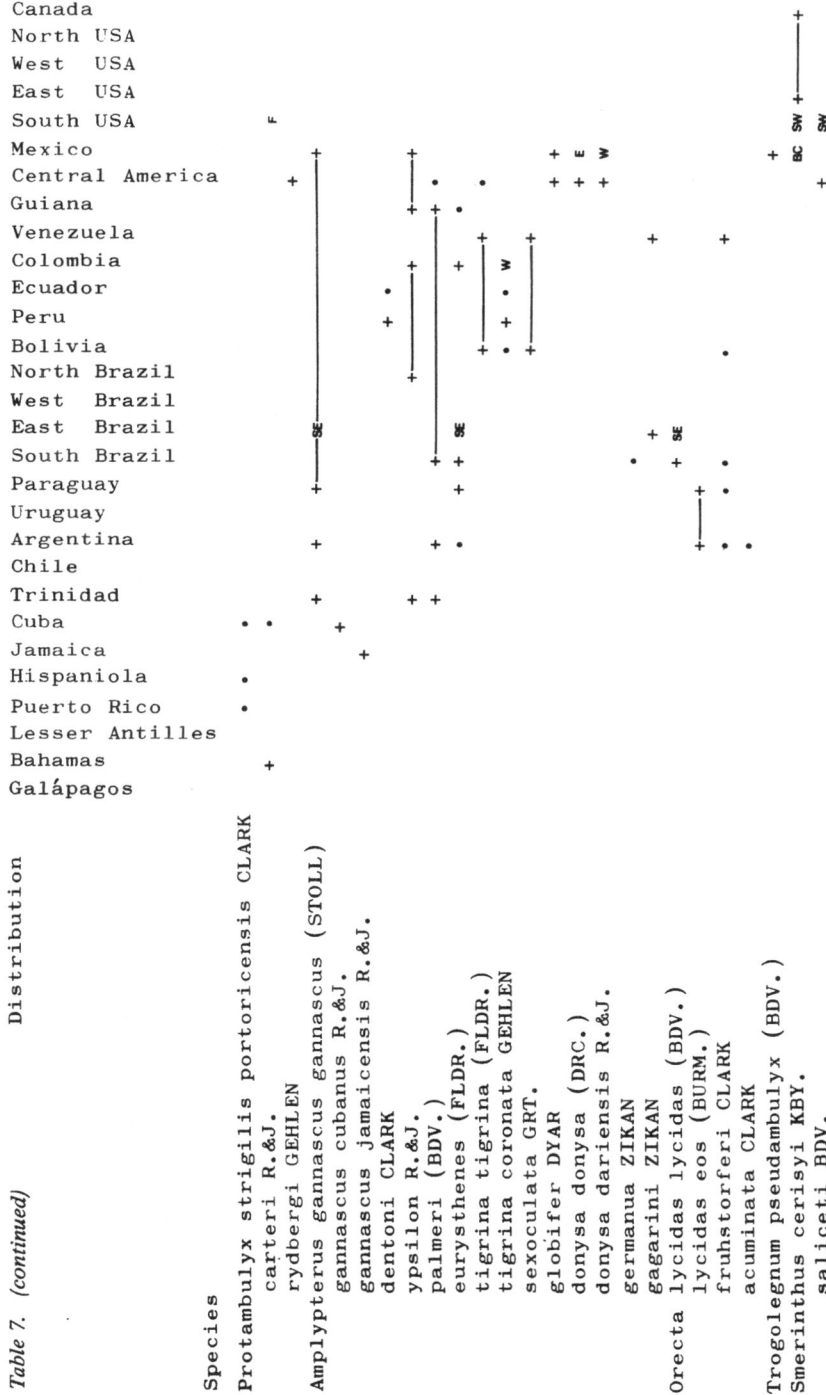

Table 7. (continued)

Species	Canada	North USA	West USA	East USA	South USA	Mexico	Central America	Guiana	Venezuela	Colombia	Ecuador	Peru	Bolivia	North Brazil	West Brazil	East Brazil	South Brazil	Paraguay	Uruguay	Argentina	Chile	Trinidad	Cuba	Jamaica	Hispaniola	Puerto Rico	Lesser Antilles	Bahamas	Galápagos
Protambulyx strigilis portoricensis CLARK																							•		•	•			
carteri R.&J.					F																		•					+	
rydbergi GEHLEN							+																						
Amplypterus gannascus gannascus (STOLL)						+	—	—	—	—	—	—	—	—	—	SE	—	+		+		+							
gannascus cubanus R.&J.																							+						
gannascus jamaicensis R.&J.																								+					
dentoni CLARK											•	+																	
ypsilon R.&J.						+	—	+		+	—	—	—	+								+							
palmeri (BDV.)							•	+	—	—	—	—	—	—	—	—	+			+		+							
eurysthenes (FLDR.)								•		+						SE	+	+		•									
tigrina tigrina (FLDR.)							•		+	—	—	—	+																
tigrina coronata GEHLEN										W	•	+	•																
sexoculata GRT.									+	—	—	—	+																
globifer DYAR						+	+																						
donysa donysa (DRC.)						E	+																						
donysa dariensis R.&J.						W	+																						
germanua ZIKAN																	•												
gagarini ZIKAN									+							+													
Orecta lycidas lycidas (BDV.)																SE	+												
lycidas eos (BURM.)																		+	—	+									
fruhstorferi CLARK									+				•				•	•		•									
acuminata CLARK																				•									
Trogolegnum pseudambulyx (BDV.)						+																							
Smerinthus cerisyi KBY.	+	—	—	+	SW	BC																							
saliceti BDV.					SW		+																						

48

Table 7. (continued)

Distribution

Species	Smerinthus jamaicensis (DRURY)	Paonias excaecatus (J.E.SMITH)	myops myops (J.E.SMITH)	myops macrops (GEHLEN)	astylus (DRURY)	Pachysphinx modesta modesta (HARR.)	modesta regalis R.&J.	modesta peninsularis CARY	occidentalis (HY.EDW.)	Monarda oryx DRC.	Cressonia juglandis (J.E.SMITH)	Pseudosphinx tetrio (L.)	Isognathus leachi (SWAINS.)	swainsoni FLDR.	scyron (STOLL)	menechus (MÉN.)	rimosus rimosus (GRT.)	rimosus inclitus EDW.	rimosus papayae BDV.	rimosus brasiliensis CLARK	rimosus wolcotti CLARK	rimosus occidentalis CLARK	rimosus molitor R.&J.	rimosus jamaicensis R.&J.	excelsior BDV.
Canada	+	+	+		+	+					+														
North USA	+				+	+					+														
West USA						+																			
East USA	+				+						+														
South USA	F	+			·	·	SW				+														
Mexico		+	+		+		BC	+			+														
Central America							·	+	·		+		+												+
Guiana							+	+			+														
Venezuela							+	+			+														
Colombia									+		·														
Ecuador																									
Peru							+	·		+													+		
Bolivia									+		+										+				
North Brazil							+	+									+								
West Brazil																		·						+	
East Brazil					SE	+	+																		
South Brazil					SE		SE				SE														
Paraguay																									
Uruguay						·																			
Argentina					+		·																		
Chile																									
Trinidad							+																		
Cuba								+																	
Jamaica													+												
Hispaniola						+	·			+															
Puerto Rico										+															
Lesser Antilles																									
Bahamas					+																				
Galápagos																									

Table 7. (continued)

Distribution

Species distribution table. Rows = geographic regions; columns = species. Marks: `+` = present, `•` = present (record), with subspecies/region codes (A, SI, F, SE, BRAZIL).

Region	tepuyensis LICHY	australis CLARK	allamandae CLARK	caricae (L.)	mossi CLARK	alope alope (DRURY)	alope dispersa KERNBACH	lassauxii (BDV.)	ello ello (L.)	ello encantada KERNBACH	yucatana (DRC.)	oenotrus (CRAM.)	crameri (SCHAUS)	obscura obscura (F.)	obscura jamaicensis CLARK	obscura conformis R.&J.	obscura socorroensis CLARK	domingonis domingonis (BUTL.)	domingonis pallescens CLARK	guttularis (WLK.)	cinifera ZIKAN	caicus (CRAM.)	ficus (L.)	syces syces (HBN.)	syces septentrionalis GEHLEN
Canada								•																	
North USA						•		+																	
West USA																				•					
East USA								+					•												
South USA				+	+			+	+A	+		+	+	•				+		+		+F	+	•	
Mexico			•						+	+A	+						SI		+				+		+
Central America			•	•					+	+A		+						•				+	+	+	+
Guiana				+																		+		+	
Venezuela	+							+									+					+		+	+
Colombia				+				+									+					+			
Ecuador																									
Peru				+				+									+					+			
Bolivia								+																	
North Brazil			+		+							+					+					+			
West Brazil									+																
East Brazil		+	+					SE																	
South Brazil	SE		SE	•				SE										+	+SE		SE BRAZIL	SE			
Paraguay															+				+				+	+	
Uruguay																									
Argentina				+				+					•				+				+		+	+	+
Chile																									
Trinidad				+	+			+				•	+	+				+				+		+	+
Cuba												+		•								+			
Jamaica														+	+										
Hispaniola										+A				+				•	•			+		•	
Puerto Rico										•								+				+			
Lesser Antilles								+														•			
Bahamas				+				+						+				+				+			
Galápagos					+	+		+									+								

49

Table 7. (continued)

Distribution

Species	Canada	North USA	West USA	East USA	South USA	Mexico	Central America	Guiana	Venezuela	Colombia	Ecuador	Peru	Bolivia	North Brazil	West Brazil	East Brazil	South Brazil	Paraguay	Uruguay	Argentina	Chile	Trinidad	Cuba	Jamaica	Hispaniola	Puerto Rico	Lesser Antilles	Bahamas	Galápagos
Pachylia syces insularis R.&J.																							+	+	+				
syces cubensis CLOSS						•	+	+		+					+	•									•				
darceta DRC.						+	+								+		+			+		+			+				
Pachylioides resumens (WLK.)					•		+	+	+				+	•	+							+							
Kloneus babayaga SKINNER							+								+							+							
Oryba kadeni (SCHAUF.)						+											SE												
achemenides CRAM.															+		SE												
Hemeroplanes triptolemus (CRAM.)									+	+	•	+		+	+					•									
diffusa (R.&J.)							P				•	+	+				SE												
ornatus ROTHSCH.											•	+					•												
longistriga (R.&J.)																	•												
Madoryx oiclus (CRAM.)						+		+	+		•	+	+	+	•		SE	+		+		+		+		•	+		
pluto pluto (CRAM.)																				+		+							
pluto dentatus GEHLEN						+	+										SE												
bubastus bubastus (CRAM.)						+	+			+		•	+	+		SE	SE	+				+					+		
bubastus butleri (KBY.)																											+		
pseudothyreus (GRT.)					F											SE	+						+					+	
Callionima nomius (WLK.)							+	+	+	+	•	+	+	+	+		•												
pan pan (CRAM.)						•	•	+	+	+	•	+					SE			+									
pan denticulata (SCHAUS)						+	+										SE			+									
pan neivae OITICICA																												+	
grisescens grisescens (ROTHSCH.)									+				+		+								+						
grisescens elegans GEHLEN										•						+							•						
gracilis (JORD.)																+													
calliommenae (SCHAUF.)						+	+	+	+	+		•										+			+				

Table 7. (continued)

Distribution

	Callionima parce parce (F.)	*parce guaycura* (CARY)	*parthenope* (ZIKAN)	*elainae* NEIDHOEFER	*falcifera* (GEHLEN)	*modesta* (GEHLEN)	*ramsdeni* CLARK	*acuta* R.&J.	*innus* (R.&J.)	*Stolidoptera tachasara* (DRC.)	*Protaleuron rhodogaster* R.&J.	*Aleuron carinata* (WLK.)	*ypanemae* (BDV.)	*cymographum* R.&J.	*chloroptera* (PERTY)	*prominens* (WLK.)	*iphis* (WLK.)	*neglectum neglectum* R.&J.	*neglectum paraguayana* CLARK	*neglectum leo* CLARK	*Pachygonia subhamata* (WLK.)	*caliginosa* (BDV.)	*hopfferi* STGR.	*martini* GEHLEN	*drucei* R.&J.
Canada																									
North USA																									
West USA																									
East USA																									
South USA	+																							+	
Mexico		BC	+				+	+		+		+		+	+			+	+		+				+
Central America			•	+			+	+		+									+	+	+	+			+
Guiana																						+	+		
Venezuela			+				+														•	•			
Colombia																					•	•			
Ecuador				•																	+				+
Peru							+		•		+									+	+				
Bolivia						+	+													+	+				+
North Brazil													•												
West Brazil		+	•				+	+		+									+		•				
East Brazil		SE																			•	•			
South Brazil							SE		+			SE		SE		SE			+	+	•	•			
Paraguay	+						+												+	+	•				
Uruguay	•											+													
Argentina	+						+							+		+									
Chile																									
Trinidad	+									+				+	+				+	+	+	+			
Cuba	+				+		•														+	+			
Jamaica		+																							
Hispaniola																									
Puerto Rico	+																								
Lesser Antilles				+																					
Bahamas																									
Galápagos																								+	

51

 Table 7. (continued)

Species	Canada	North USA	West USA	East USA	South USA	Mexico	Central America	Guiana	Venezuela	Colombia	Ecuador	Peru	Bolivia	North Brazil	West Brazil	East Brazil	South Brazil	Paraguay	Uruguay	Argentina	Chile	Trinidad	Cuba	Jamaica	Hispaniola	Puerto Rico	Lesser Antilles	Bahamas	Galápagos
Pachygonia ribbei ribbei DRC.							+		+																				
ribbei peruviana J.&T.											•	+																	
Nyceryx hyposticta (FLDR.)									+	—	—	—	+																
ericea (DRC.)							+		+	—	—	—	+		+		•												
lunaris JORD.							•				+																		
coffeae (WLK.)					F	+	—	—	—	—	—	—	—	—	+		SE					+							
magna (FLDR.)					•					•	+	+		+	+														
tacita (DRC.)						+	+	•	•		+	+	•			SE													
eximia R.&J.							+			•		•																	
maxwelli (ROTHSCH.)									+		+	—	+			•	+												
nictitans nictitans (BDV.)							•			•	•					SE	+			•									
nictitans saturata R.&J.										•	+	—	+																
continua continua (WLK.)																SE	+			•									
continua cratera R.&J.												+	+																
alophus alophus (BDV.)										•						SE	+	•		+									
alophus ixion (BURM.)																	+	—	—	+									
lemoni GEHLEN																S. AMERICA													
nephus (BDV.)																SE													
riscus (SCHAUS)						+	—	—	—	—	—	—	—	—	+	•	SF	+				+							
stuarti (ROTHSCH.)					•			+	+		+	—	—	—	—	SE	•												
draudti GEHLEN												+																	
müllerì CLARK						+																							
clarki FASSL												•	+																
Perigonia divisa G.&R.																							+						
grisea R.&J.											+	—	+																

Table 7. *(continued)*

Distribution

Canada
North USA
West USA
East USA
South USA
Mexico
Central America
Guiana
Venezuela
Colombia
Ecuador
Peru
Bolivia
North Brazil
West Brazil
East Brazil
South Brazil
Paraguay
Uruguay
Argentina
Chile
Trinidad
Cuba
Jamaica
Hispaniola
Puerto Rico
Lesser Antilles
Bahamas
Galápagos

Species

Perigonia pallida pallida R.&J.
pallida rufescens DANIEL
thayeri CLARK
pittieri LICHY
stulta H.-S.
leucopus R.&J.
lusca lusca (F.)
lusca passerina R.&J.
lusca ilus BDV.
lusca continua VAZQUEZ
interrupta WLK.
lefeburei (LUC.)
manni CLARK
jamaicensis ROTHSCH.
glaucescens WLK.
Eupyrrhoglossum sagra (POEY)
corvus (BDV.)
venustum R.&J.
Aellopos ceculus (CRAM.)
gehleni (CLOSS)
blaini H.-S.
tantalus (L.)
clavipes clavipes (R.&J.)
clavipes eumelas (JORD.)
titan titan (CRAM.)

Table 7. *(continued)*

54

Distribution

Species	Distribution (regions with records)
Aellopos titan cubana (CLARK)	Colombia •, Cuba +
titan aguacana (GEHLEN)	Mexico +, South Brazil •, Paraguay +, Trinidad +, Cuba +, Hispaniola +
fadus fadus (CRAM.)	South USA (F), Mexico +, West Brazil +, South Brazil (SE), Paraguay +, Trinidad +
fadus flavosignata (CLOSS)	Mexico +, Ecuador +, Bolivia +, East Brazil +, South Brazil (SE), Trinidad +, Jamaica •
Enyo japix japix (CRAM.)	East USA +, Mexico +, Trinidad +, Cuba +, Jamaica •, Hispaniola +, Bahamas +, Galápagos +
japix discrepans (WLK.)	Guiana +, Argentina +, Trinidad +, Lesser Antilles +
pronoe pronoe (DRC.)	South USA •, Mexico +, Cuba •, Jamaica •
pronoe fuscatus R.&J.	—
lugubris lugubris (L.)	South USA (F), Paraguay +, Trinidad +, Cuba •, Hispaniola +
lugubris delanoi (KERNBACH)	Mexico +, Paraguay +
lugubris latipennis (R.&J.)	Mexico •
ocypete (L.)	—
boisduvalii (OBTH.)	Paraguay +, Argentina +, Trinidad +, Cuba +, Hispaniola +
gorgon gorgon (CRAM.)	Mexico •, Central America +, Venezuela +, Colombia +, Peru •, Bolivia +, Paraguay +, Argentina +, Trinidad +, Cuba •
gorgon heinrichi CLARK	Mexico +, Central America + (P)
taedium (SCHAUS)	Central America +
bathus bathus (ROTHSCH.)	Mexico +, Peru •, Bolivia +
bathus otiosus (KERNBACH)	Ecuador •, Peru +, Bolivia +
cavifer cavifer (R.&J.)	Mexico +, Colombia +, Bolivia •, East Brazil (SE), South Brazil •
cavifer reconditus (KERNBACH)	Mexico •, North Brazil +, South Brazil +
cavifer paganus (KERNBACH)	—
Hemaris thysbe (F.)	Canada +, North USA +, East USA +, South USA +
gracilis (G.&R.)	Canada +, North USA + (NE)
diffinis (BDV.)	Canada +, North USA + (NW), South USA + (BC)
senta (STKR.)	Canada +, North USA +

Table 7. (continued)

Distribution

Species

Region	anchemola	triangulum	satellitia satellitia	satellitia analis	satellitia excessa	satellitia posticata	pandorus	macasensis	drucei	neuburgeri	elisa	cissi	obliquus obliquus	obliquus orientalis	eacus	adamsi	translineatus	achemon	typhon	strenua	vitis vitis	vitis hesperidum	vitis fuscatus	fasciata fasciata	fasciata tupaci
Canada							+																		
North USA							NE																		E
West USA																		+							
East USA							+											+							
South USA	.	+	T			F	+	+								.		+ A							
Mexico	+	+					+					+			+			+ A						+	
Central America		+							.				+		+			+ H					.		
Guiana					.																				
Venezuela		+							.				+		+		+								
Colombia											+														
Ecuador											+	+			+										
Peru												.			+										
Bolivia		+		+				.					+	+	+										
North Brazil															+										
West Brazil			+	+																					
East Brazil			SE										SE		SE	SE	SE								
South Brazil	+	.	+										+		+	+	+								
Paraguay															+	.									
Uruguay	.								.																
Argentina	+	+	+										+					+			.				
Chile																									
Trinidad	+	+											+								+			+	
Cuba		.	+										+								+	+			
Jamaica		+		.																		+			
Hispaniola																	.				+	+			
Puerto Rico																						+			
Lesser Antilles													+											+	+
Bahamas								+													+				+
Galápagos																									+

Eumorpha anchemola (CRAM.)
triangulum (R.&J.)
satellitia satellitia (L.)
satellitia analis (R.&J.)
satellitia excessa (GEHLEN)
satellitia posticata (GRT.)
pandorus (HBN.)
macasensis (CLARK)
drucei (R.&J.)
neuburgeri (R.&J.)
elisa (SMYTH)
cissi (SCHAUF.)
obliquus obliquus (R.&J.)
obliquus orientalis (DANIEL)
eacus (CRAM.)
adamsi (R.&J.)
translineatus (ROTHSCH.)
achemon (DRURY)
typhon (KLUG)
strenua (MÉN.)
vitis vitis (L.)
vitis hesperidum (KBY.)
vitis fuscatus (R.&J.)
fasciata fasciata (SULZ.)
fasciata tupaci (KERNBACH)

Table 7. (continued)

Distribution

Species	Canada	North USA	West USA	East USA	South USA	Mexico	Central America	Guiana	Venezuela	Colombia	Ecuador	Peru	Bolivia	North Brazil	West Brazil	East Brazil	South Brazil	Paraguay	Uruguay	Argentina	Chile	Trinidad	Cuba	Jamaica	Hispaniola	Puerto Rico	Lesser Antilles	Bahamas	Galápagos
Eumorpha phorbas (CRAM.)							+							+		+				·		+						+	+
capronnieri (BDV.)							+							+		·						+							
labruscae labruscae (L.)	+	+	·		+											SE												+	
labruscae yupanquii (KERNBACH)																													
Himantoides undata (WLK.)								·	·																				
perkinsae CLARK							·																						
Cautethia spuria (BDV.)					T	+	+																GC	+	+			+	
yucatana CLARK					T	+	+																LC						
grotei grotei HY.EDW.					F	T																		+					
grotei apira JORD.						T																		+					
grotei hilaris JORD.						F																							
noctuiformis noctuiformis (WLK.)																							+		+		SJ		
noctuiformis bredini CARY																											N		
simitia SCHAUS							·			+																			
Gurelca muelleri CLARK						W																							
sonorensis CLARK						W																							
Sphecodina abbottii (SWAINS.)	+	NE			F																								
Deidamia inscripta (HARR.)	+	NE		+	F																								
Arctonotus lucidus BDV.			A																										
Amphion nessus (CRAM.)	+			A	+	+	+																						
Proserpinus gaurae (J.E.SMITH)		NW			+	+	+																						
juanita (STKR.)					SE																								
clarkiae (BDV.)			NW		+		BC																						
flavofasciata (WLK.)	+	+	NW	+	+	+																							
vega (DYAR)	+	+			NE	+	+																						

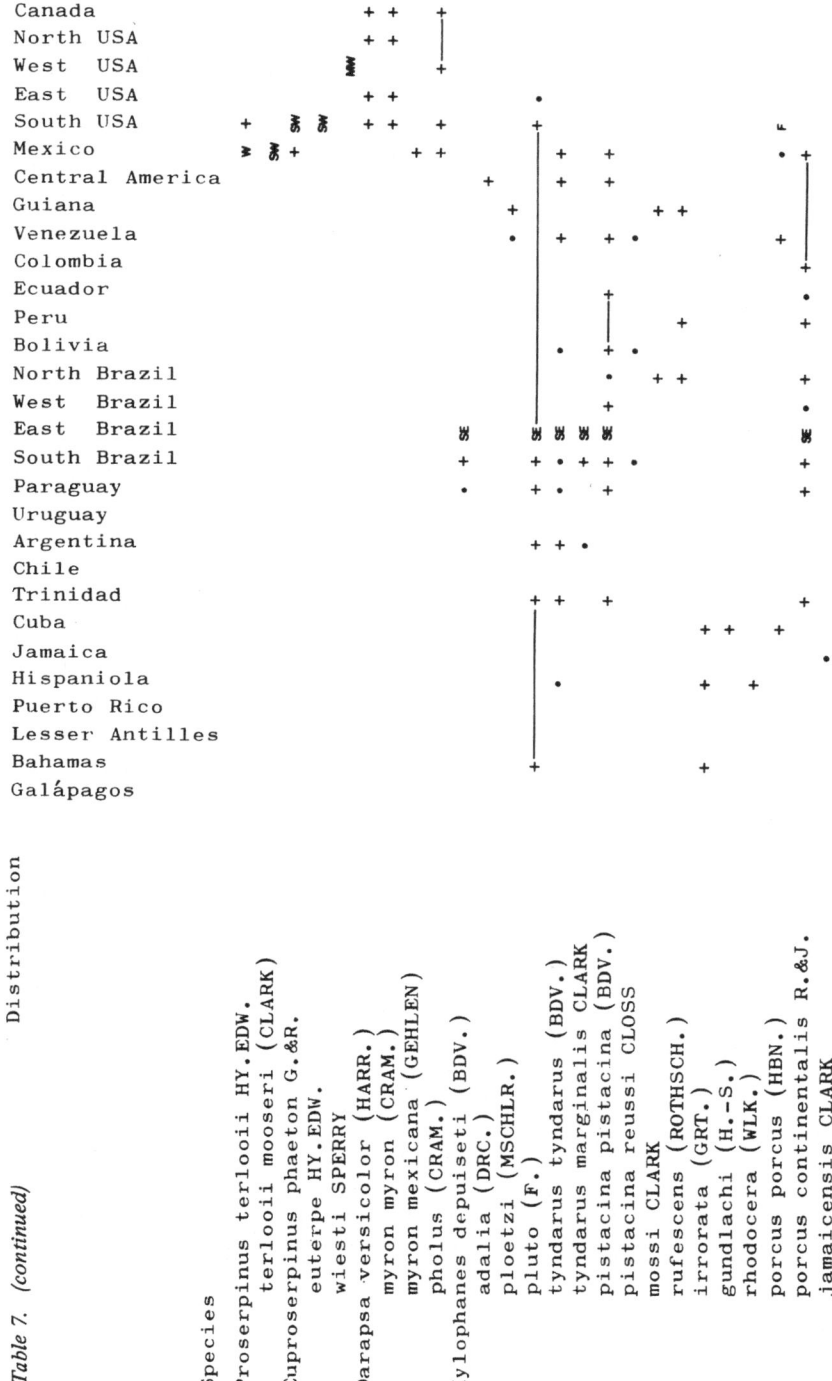

Canada
North USA
West USA
East USA
South USA
Mexico
Central America
Guiana
Venezuela
Colombia
Ecuador
Peru
Bolivia
North Brazil
West Brazil
East Brazil
South Brazil
Paraguay
Uruguay
Argentina
Chile
Trinidad
Cuba
Jamaica
Hispaniola
Puerto Rico
Lesser Antilles
Bahamas
Galápagos

Distribution

Table 7. (continued)

Species
Proserpinus terlooii HY.EDW.
  terlooii mooseri (CLARK)
Euproserpinus phaeton G.&R.
  euterpe HY.EDW.
  wiesti SPERRY
Darapsa versicolor (HARR.)
  myron myron (CRAM.)
  myron mexicana (GEHLEN)
  pholus (CRAM.)
Xylophanes depuiseti (BDV.)
  adalia (DRC.)
  ploetzi (MSCHLR.)
  pluto (F.)
  tyndarus tyndarus (BDV.)
  tyndarus marginalis CLARK
  pistacina pistacina (BDV.)
  pistacina reussi CLOSS
  mossi CLARK
  rufescens (ROTHSCH.)
  irrorata (GRT.)
  gundlachi (H.-S.)
  rhodocera (WLK.)
  porcus porcus (HBN.)
  porcus continentalis R.&J.
  jamaicensis CLARK

57

Table 7. (continued)

Species	Canada	North USA	West USA	East USA	South USA	Mexico	Central America	Guiana	Venezuela	Colombia	Ecuador	Peru	Bolivia	North Brazil	West Brazil	East Brazil	South Brazil	Paraguay	Uruguay	Argentina	Chile	Trinidad	Cuba	Jamaica	Hispaniola	Puerto Rico	Lesser Antilles	Bahamas	Galápagos
Xylophanes hannemanni CLOSS						+																							
schausi schausi (ROTHSCH.)																SE	+												
schausi serenus R.&J.							P		+				+	+															
germen germen (SCHAUS)					•	+	+		+	—	—	—	+																
germen yurakano LICHY									+																				
hojeda GEHLEN										W		•																	
brevis CLARK						+	+		+	+																			
juanita R.&J.						+	+			•							•												
fusimacula fusimacula (FLDR.)								+	—	+	•	+	+		•														
fusimacula niepelti GEHLEN										W	•																		
zurcheri (DRC.)						•	+			+																			
undata R.&J.							+			+	—	—	—	NW															
rhodina R.&J.							+																						
godmani (DRC.)							P					•																	
falco (WLK.)					+	+	H																						
xylobotes (BURM.)												+	+	•		SE	+			+									
media R.&J.								+	+			+	+																
ceratomioides (G.&R.)						+	—	—	—	—	—	—	+			SF	+	+		•		+	•						
guianensis (ROTHSCH.)								+	+		+	+		+			•												
anubus anubus (CRAM.)						+	—	—	—	—	—	—	—	+		SE	+	+		+		+							
anubus paraguayensis GEHLEN																		+											
docilis (BUTL.)							•				+	—	+																
amadis amadis (STOLL)						•	•	+	•										I										
amadis meridianus R.&J.							•		+	+																			
amadis cyrene (DRC.)						+	+			•																			

Table 7. (continued)   Distribution

Columns (left to right): Galápagos, Bahamas, Lesser Antilles, Puerto Rico, Hispaniola, Jamaica, Cuba, Trinidad, Chile, Argentina, Uruguay, Paraguay, South Brazil, East Brazil, West Brazil, North Brazil, Bolivia, Peru, Ecuador, Colombia, Venezuela, Guiana, Central America, Mexico, South USA, East USA, West USA, North USA, Canada

Species	Galápagos	Bahamas	Lesser Antilles	Puerto Rico	Hispaniola	Jamaica	Cuba	Trinidad	Chile	Argentina	Uruguay	Paraguay	South Brazil	East Brazil	West Brazil	North Brazil	Bolivia	Peru	Ecuador	Colombia	Venezuela	Guiana	Central America	Mexico	South USA	East USA	West USA	North USA	Canada
Xylophanes amadis stuarti (ROTHSCH.)																		+——+											
amadis goeldi R.&J.																	SE	+	•	+			+						
epaphus (BDV.)																			+			+							
acrus R.&J.													+																
alegrensis CLOSS																		•											
agilis CLOSS																							+	+					
belti (DRC.)																	•	+	+										
rothschildi rothschildi (DOGN.)																	+	•	•										
rothschildi fassli GEHLEN																		+											
rothschildi bilineata GEHLEN																		+											
cosmius cosmius R.&J.														+	+														
cosmius obscurus R.&J.																	+												
ockendeni ROTHSCH.																		•	+										
mirabilis CLARK																		+											
macasensis CLARK			•	•	+	+																							
chiron chiron DRURY																													•
chiron nechus (CRAM.)							•	•	+		•	•	+	+————————————————+															
chiron martiniquensis KERNBACH			M																										
chiron lucianus R.&J.			+																										
chiron cubanus R.&J.		+					+																						
crotonis (WLK.)														+					+————————+	+								
aristor (BDV.)												•	•					+————————+	+									
schreiteri CLARK										+																			
rhodochlora R.&J.													+	+		+	•												
eumedon (BDV.)														•			•	•	+	G	+								

Table 7. (continued)

Distribution

	rhodotus	caissa	nabuchodonosor	titana	indistincta	muelleri	resta	tersa tersa	tersa cubensis	norfolki	suana	turbata	fosteri	clarki	ferotinus	dolius	kaempferi	colombiana	elara elara	elara simulans	isaon isaon	isaon nanus	hydrata	robinsoni	josephinae
Canada								•	•																
North USA								+																	
West USA																									
East USA								+																	
South USA																									
Mexico				+	+			•				+													
Central America			•			•		•				+													G
Guiana																	•		+						
Venezuela					+			+				+		+					+						
Colombia			•					+				•						+							
Ecuador	•															+									
Peru	+		+					+								+			+					•	
Bolivia	•	+	+	+				+								+			•	+					
North Brazil															NW				+						
West Brazil																									
East Brazil			SE		SE							SE			SE			SE		SE		SE			
South Brazil			+		+											+	+		SW	+	SW	+			
Paraguay																+		+		+					
Uruguay																									
Argentina								+				+													
Chile								+																	
Trinidad								+					+												
Cuba								•	+						+									+	
Jamaica								+																	
Hispaniola																									
Puerto Rico																									
Lesser Antilles								+																	
Bahamas										+															
Galápagos								+	+																

Species

Xylophanes rhodotus ROTHSCH.
caissa GEHLEN
nabuchodonosor OBERTH.
titana (DRC.)
indistincta CLOSS
muelleri CLARK
resta R.&J.
tersa tersa (L.)
tersa cubensis GEHLEN
norfolki KERNBACH
suana (DRC.)
turbata (HY.EDW.)
fosteri R.&J.
clarki RAMSDEN
ferotinus GEHLEN
dolius R.&J.
kaempferi CLARK
colombiana CLARK
elara elara (DRC.)
elara simulans RAYMUNDO
isaon isaon (BDV.)
isaon nanus RAYMUNDO
hydrata R.&J.
robinsoni (GRT.)
josephinae CLARK

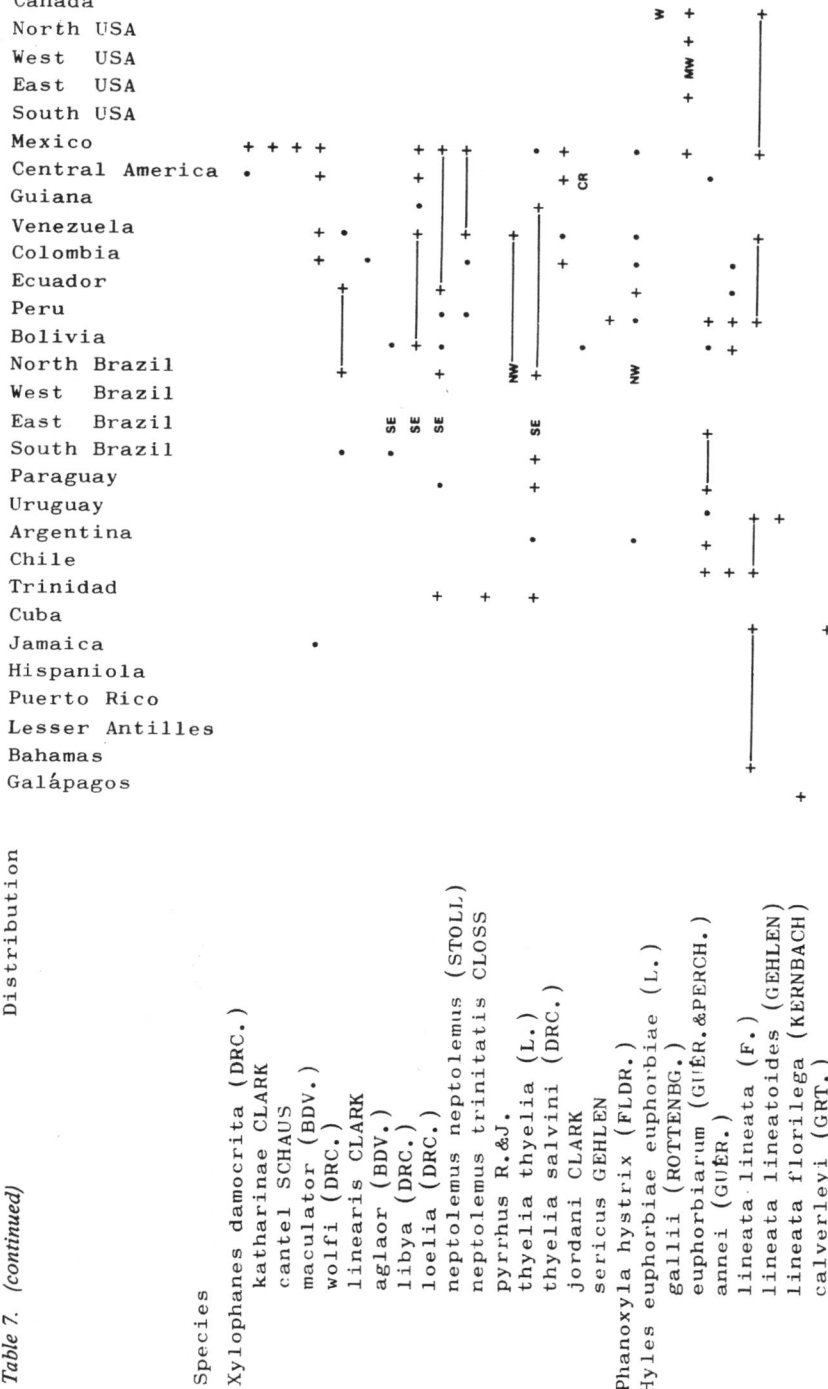

Table 7. (continued)

Distribution

Canada
North USA
West USA
East USA
South USA
Mexico
Central America
Guiana
Venezuela
Colombia
Ecuador
Peru
Bolivia
North Brazil
West Brazil
East Brazil
South Brazil
Paraguay
Uruguay
Argentina
Chile
Trinidad
Cuba
Jamaica
Hispaniola
Puerto Rico
Lesser Antilles
Bahamas
Galápagos

Species
Xylophanes damocrita (DRC.)
 katharinae CLARK
 cantel SCHAUS
 maculator (BDV.)
 wolfi (DRC.)
 linearis CLARK
 aglaor (BDV.)
 libya (DRC.)
 loelia (DRC.)
 neptolemus neptolemus (STOLL.)
 neptolemus trinitatis CLOSS
 pyrrhus R.&J.
 thyelia thyelia (L.)
 thyelia salvini (DRC.)
 jordani CLARK
 sericus GEHLEN
Phanoxyla hystrix (FLDR.)
Hyles euphorbiae euphorbiae (L.)
 gallii (ROTTENBG.)
 euphorbiarum (GUÉR.&PERCH.)
 annei (GUÉR.)
 lineata lineata (F.)
 lineata lineatoides (GEHLEN)
 lineata florilega (KERNBACH)
 calverleyi (GRT.)

61

Table 8. List of Sphingidae of the New World with distributions other than exclusively Neotropical: I Nearctic sphingids; II Nearctic sphingids which, with the exception of *Hyles lineata*, are known to occur only sporadically also in the Neotropical north, III Neotropic-Nearctic sphingids, IV Neotropical sphingids with sporadic occurrence in the Nearctic region.* Palaearctic or Palaearctic-Nearctic

I	II	III	IV
1 Manduca rustica cortesi	Manduca sexta sexta	Agrius cingulatus	Cocytius duponchel
2 Manduca jasminearum	Manduca quinquemaculata	Cocytius antaeus medor	Cocytius antaeus antaeus
3 Dolba hyloeus	Paratrea plebeja	Manduca rustica rustica	Neococytius cluentius
4 Ceratomia hageni	Smerinthus saliceti	Manduca brontes cubensis	Manduca occulta occulta
5 Ceratomia amyntor	Pachysphinx occidentalis	Manduca florestan florestan	Manduca pellenia pellenia
6 Ceratomia undulosa	Eumorpha pandorus	Sphinx istar	Manduca ochus
7 Ceratomia sonorensis	Eumorpha achemon	Protambulyx carteri	Manduca albiplaga albiplaga
8 Ceratomia catalpae	Amphion nessus	Pseudosphinx tetrio	Manduca muscosa
9 Isoparce cupressi	Proserpinus juanita	Erinnyis alope alope	Dolbogene hartwegii
10 Sagenosoma elsa	Proserpinus vega	Erinnyis ello ello	Sphinx lugens
11 Sphinx eremitus	Proserpinus terlooii terlooii	Erinnyis crameri	Sphinx geminus
12 Sphinx eremitoides	Darapsa pholus	Erinnyis obscura obscura	Sphinx separata melaena
13 Sphinx separata separata	Hyles lineata lineata	Erinnyis domingonis domingonis	Sphinx chisoya
14 Sphinx chersis chersis		Pachylia ficus	Sphinx leucophaeata
15 Sphinx vashti		Madoryx pseudothyreus	Protambulyx strigilis strigilis
16 Sphinx libocedrus libocedrus		Callionima parce parce	Trogolegnum pseudambulyx
17 Sphinx perelegans		Perigonia lusca lusca	Isognathus rimosus inclitus
18 Sphinx asella		Aellopos tantalus	Erinnyis lassauxii
19 Sphinx canadensis		Aellopos clavipes clavipes	Erinnyis yucatana
20 Sphinx franckii		Aellopos titan titan	Erinnyis oenotrus
21 Sphinx kalmiae		Aellopos fadus fadus	Erinnyis guttularis
22 Sphinx gordius		Enyo lugubris lugubris	Phryxus caicus
23 Sphinx luscitiosa		Eumorpha vitis vitis	Pachylia syces syces
24 Sphinx drupiferarum		Eumorpha fasciata fasciata	Pachylioides resumens
25 Sphinx dollii		Eumorpha labruscae labruscae	Callionima inuus
26 Sphinx sequoiae		Cautethia grotei grotei	Pachygonia caliginosa
27 Sphinx pinastri pinastri*		Xylophanes pluto	Nyceryx coffeae

Table 8. *(continued)*

I	III	IV
28 Lapara coniferarum	Xylophanes falco	Nyceryx magna
29 Lapara bombycoides	Xylophanes tersa tersa	Nyceryx stuarti
30 Smerinthus cerisyi		Enyo ocypete
31 Smerinthus jamaicensis		Enyo gorgon gorgon
32 Paonias excaecatus		Eumorpha anchemola
33 Paonias myops myops		Eumorpha triangulum
34 Paonias myops macrops		Eumorpha satellitia satellitia
35 Paonias astylus		Eumorpha satellitia posticata
36 Pachysphinx modesta modesta		Eumorpha eacus
37 Pachysphinx modesta peninsularis		Eumorpha typhon
38 Monarda oryx		Cautethia spuria
39 Cressonia juglandis		Cautethia yucatana
40 Callionima parce guaycura		Xylophanes porcus porcus
41 Hemaris thysbe		Xylophanes germen germen
42 Hemaris gracilis		
43 Hemaris diffinis		
44 Hemaris senta		
45 Gurelca sonorensis		
46 Sphecodina abbottii		
47 Deidamia inscripta		
48 Arctonotus lucidus		
49 Proserpinus gaurae		
50 Proserpinus clarkiae		
51 Proserpinus flavofasciata		
52 Euproserpinus phaeton		
53 Euproserpinus euterpe		
54 Euproserpinus wiesti		
55 Darapsa versicolor		
56 Darapsa myron myron		
57 Hyles euphorbiae euphorbiae*		
58 Hyles gallii*		

c. Distribution of sphingid species and subspecies and distribution centres established

In searching to illustrate the distribution of the Sphingidae within the Neotropical region in such a way that both the falling gradient towards the north and the extreme south and the approximate position of centres of concentration could be seen, I came across the illustration technique used by ERWIN (1970). He refers to a zoogeographical analysis by BALL & FREITAG (in FREITAG 1969). However, the method had first been used by SIMPSON for North American mammals (SIMPSON, 1964).

Comparable to the technique used in the present work, the total number of species occurring in a grid field as obtained by overlaying dot maps with a grid

63

map, is determined by plotting all species ranges on a map of the region. The grid fields are artificial with regard to their borders and individual character, it is true, but they have the advantage of being defined exactly by selected longitudes and latitudes. Thus, the data given in fig. 7 can be checked at any time and revised at a later date. This will certainly be necessary, since, so far, the species number per unit is partly governed by the collecting activities in the area, and partly the state of knowledge regarding the validity of what is presently considered as species and subspecies.

Specimens clearly identified as strays have not been included in the numbers given. For areas without numbers, the exact number of species has not been figured out. As against this, areas, for which no records were known, have been marked with the number zero.

With regard to the share some areas have in the ocean, I desisted from the possibility to use, like ERWIN (1970), a parameter which he calls ALIV (Average Landmass Interval Value). Instead, my figures, which anyway represent only approximate values, refer, as shown on the example of area J4, to the number of species found in the whole of Florida or, like in the case of K6, to the number of species found on Jamaica excluding the Cuban share in this field.

Besides 432 North and South American sphingid species 144 additional subspecies are known (cf. table 7). Subspecies are of particular importance in chorological investigations. Altogether, 435 specifically and subspecifically differentiated Sphingidae have an exclusively Neotropical distribution, as against 58* from the Nearctic region (cf. table 8). Another 13 Nearctic sphingids have also been found in the Neotropical north. In comparison, a total of 29 Neotropical sphingids have a share in the Nearctic region. Another 41 Neotropical sphingids occur sporadically in the Nearctic region.

For the establishment of centres of dispersal, the mapped distribution ranges of all 518 sphingid species and subspecies occurring parly or wholly in the Neotropical region have been evaluated. In doing so, taxa not yet fully confirmed and, hence, with an uncertain status, have been handled with just as much caution as widespread polycentric species. In the latter case, the phylogeny of the subunits and their centres of differentiation and distribution have first to be investigated, before it can be decided whether polycentric species pertain to certain centres as their faunal elements. An especially great power of evidence can be expected from monotypic species with small ranges. Their distribution ranges, if plotted one above the other on a map of the region, lead to areas of approximate congruence, the so-called range centres or 'nuclear areas' (REINIG 1937, 1950, DE LATTIN 1957, MÜLLER 1970, 1971, 1973).

By applying this method on the basis of the mapped distribution of Neotropical sphingids, 18 centres could be established (fig. 8). In naming them, I followed, whenever possible, the designation chosen by MÜLLER (1970) (cf. fig. 9).

The mentioned centres will first be looked upon and verified as centres of distribution.

* including two Palaearctic and one Palaearctic-Nearctic species

64

Popular objects of zoogeographical investigations are islands, because they exhibit the isolation conditions necessary for speciation also in the present (DE LATTIN 1967). The study of island faunas led DARWIN (1859) to valuable findings. Recently, it has especially been MAC ARTHUR & WILSON (1967) who dealt with the biogeographical importance of islands.

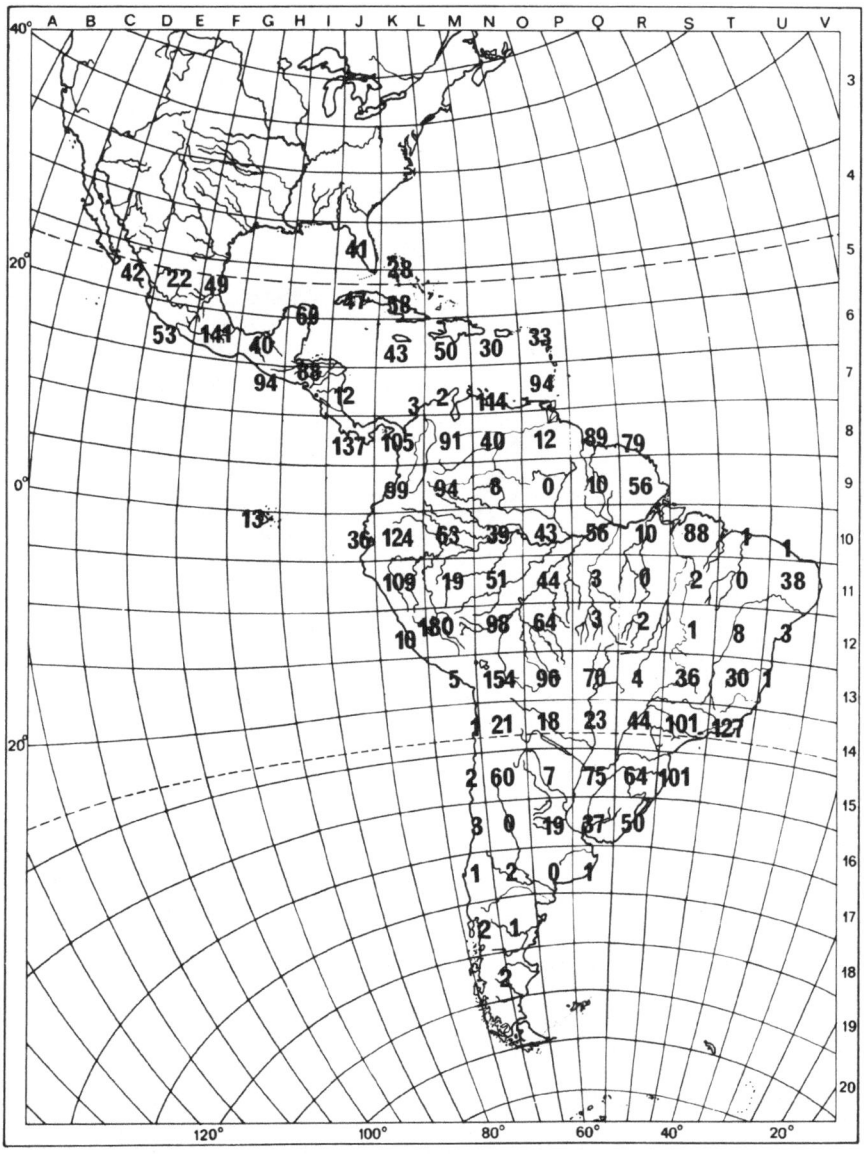

Fig. 7. Distribution of the species and subspecies of Sphingidae in the investigated area.

65

I would like to discuss first the centres of the Galápagos and the West Indian Islands and then try to reproduce the 'insular conditions' on the continent as they have been prevailing there at least previously and inasmuch as they have been important for the sphingids.

Fig. 8. Dispersal centres as established with Neotropical Sphingidae: 1 = Galápagos centre; 2 = East Cuban centre; 3 = Hispaniola centre; 4 = Jamaica centre; 5 = Lesser Antilles centre; 6 = Central American rain-forest centre; 7 = Central American Pacific centre; 8 = Yucatán centre; 9 = Costa Rican centre; 10 = Yungas centre; 11 = Tucuman centre; 12 = Venezuelan centre; 13 = Serra do Mar centre; 14 = Uruguayan centre; 15 = Andean Pacific centre; 16 = Paraguayan centre; 17 = Mato Grosso centre; 18 = Guiana centre.

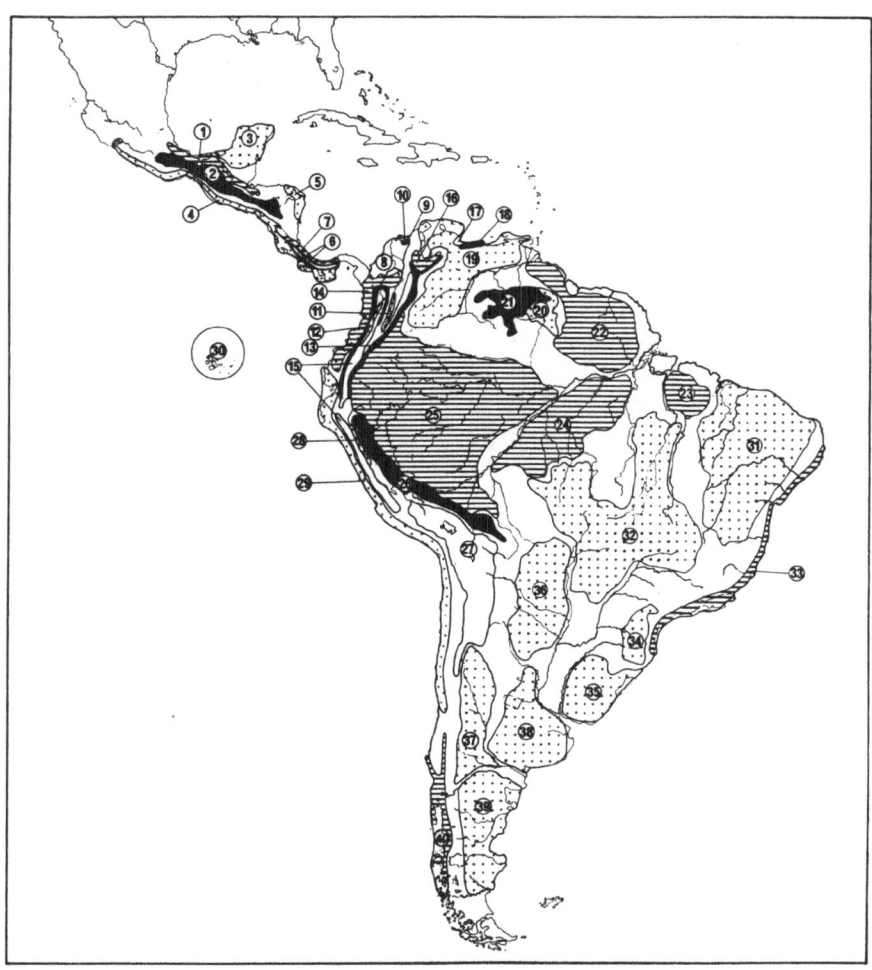

Fig. 9. 'The dispersal centres of terrestrial vertebrates in the Neotropical realm' as analysed by MÜLLER (1970). Black = montane forest centres; diagonal hatching = rainforest centres; white = oreal centres; stippling = non-forest centres; 1 = Central American rain-forest centre; 2 = Central American montane forest centre; 3 = Yucatan centre; 4 = Central American Pacific centre; 5 = Coco centre; 6 = Costa Rican centre; 7 = Talamanca paramo centre; 8 = Barranquilla centre; 9 = Santa Marta centre; 10 = Sierra Nevada centre; 11 = Magdalena centre; 12 = Cauca centre; 13 = Colombian montane forest centre; 14 = Colombian Pacific centre; 15 = North Andean centre; 16 = Catatumbo centre; 17 = Venezuelan coastal forest centre; 18 = Venezuelan montane forest centre; 19 = Caribbean centre; 20 = Roraima centre; 21 = Pantepui centre; 22 = Guyanan centre; 23 = Para centre; 24 = Madeira centre; 25 = Amazon centre; 26 = Yungas centre; 27 = Puna centre; 28 = Marañon centre; 29 = Andean Pacific centre; 30 = Galapagos centre; 31 = Caatinga centre; 32 = Campo Cerrado centre; 33 = Serra do Mar centre; 34 = Parana centre; 35 = Uruguayan centre; 36 = Chaco centre; 37 = Monte centre; 38 = Pampa centre; 39 = Patagonian centre; 40 = Nothofagus centre.

67

1. The Galápagos centre

The Galápagos archipelago which is situated 950 km west of the coast of Ecuador, consists of 9 larger and more than 40 smaller islands covering a total area of 7844 km². They are of volcanic origin, and in the west they reach maximum heights of more than 1500 m. In spite of their position on the equator, their climate is described as moderate and dry. The lowlands are said to have almost desert character (MAULL and others 1930), the vegetation in the interior is, however, abundant. The forest cover of the mountains which reaches further down on the southeast slopes, changes, at a height of about 200 m, into meagre thornbush and reaches the height of 600 m as cloud forest. Then follows, on the more hilly islands (according to MAULL and others 1930), grassland comparable to the paramos of the Andes in five to six times higher elevations.

Many authors have been working on the Galápagos archipelago and its entomofauna, since these islands had been brought into focus by DARWIN's observations (BEEBE 1924, BOWMAN 1966, CLARK 1926, CURIO 1965, DENBURGH & SLEVIN 1913, EIBL-EIBESFELDT 1959, 1960, HAYES 1975, KERNBACH 1962b and 1964a, DE LATTIN 1967, LINELL 1899, LINDSLEY & USINGER 1966, MELVILLE 1946, MORGAN 1920, WHEELER 1919, WILLIAMS 1911 and 1926).

MÜLLER (1970, 1973) says that the inventory of the plants and animals of this group of islands can be taken as essentially complete. This statement is supported by the investigations of LINSLEY & USINGER (1966), who could name 618 insect species as occurring on the Galápagos Islands (among them 97 Lepidoptera). This amazingly low number − it is lower than the established number of plant species (700) − is explained by the scanty occurrence of freshwater. MAULL and others (1930) say that, even in higher elevations, running water is rare due to the permeable volcanic rocks. This observation is substantiated by the lack of animal groups like amphibians and freshwater fishes (MÜLLER 1970, 1973).

The poverty in insect species is further explained by the distance from the continent. A successful colonization by active or passive immigrants is not likely to take place easily, especially because of the high pressure of selection (CURIO 1965). On the other hand, the distance from the continent created favourable isolation conditions, which is reflected by the richness in endemics. On the basis of the studies by WILLIAMS (1911), CLARK (1926) and the more recent works of KERNBACH (1962b, 1964a) and HAYES (1975) the following 13 spingid species and subspecies can be listed for the Galápagos Islands, of which as many as 10 are endemics (e) differentiated in one species and 9 subspecies:

	1.	Agrius cingulatus
(e)	2.	Manduca sexta leucoptera
(e)	3.	Manduca rustica calapagensis
(e)	4.	Erinnyis alope dispersa
(e)	5.	Erinnyis ello encantada

(e) 6. Erinnyis obscura conformis
(e) 7. Enyo lugubris delanoi
     8. Pachygonia drucei
(e) 9. Eumorpha fasciata tupaci
(e) 10. Eumorpha labruscae yupanquii
    11. Xylophanes tersa
(e) 12. Xylophanes norfolki
(e) 13. Hyles lineata florilega

All endemics are derived from species widely spread on the mainland. Almost all sphingids which have been shown to occur on the Galápagos Islands are smaller than on the continent (CURIO 1965) and also less colourful (DARWIN 1859).

The subspecies *calapagensis* of *Manduca rustica* has been considered as a distinct species by CLARK (1926) and DRAUDT (1931), whereas ROTHSCHILD & JORDAN (1903) and KERNBACH (1962b) looked upon it as a subspecies. DRAUDT (1931) listed *Manduca nigrita* as a distinct species from the Galápagos Islands, whereas KERNBACH (1962b) considered it to be an aberration only of *Manduca rustica calapagensis*.

On 9 of the Galápagos Islands endemic sphingids have been recorded (fig. 10). Of the three non-endemic species, *Agrius cingulatus* is known to occur on Isabela, San Cristóbal, Santa Cruz, Santa Maria and Baltra, *Xylophanes tersa* on San Cristóbal and Santa Maria; of *Pachygonia drucei* one specimen in the collection of the British Museum is labelled 'Galápagos & Cocos Islands' (doubtful record according to HAYES 1975).

It has not yet been established whether the 11 sphingid species occurring on Santa Cruz (including *Agrius cingulatus*) would characterize this island as a natural centre of species diversity within the Galápagos archipelago. On the basis of the existing endemic sphingids alone it can, however, be maintained that the Galápagos archipelago represents a definite individual zoogeographic unit which allows us to refer to it as the Galápagos centre.

The centres of the West Indian Islands

With regard to zoogeographical conditions, the West Indian Islands differ from the Galápagos Islands mainly by their greater land-masses (234,732 km^2) and their closeness to the mainland. Cuba is separated from the southernmost point of Florida by the only 180 km wide Straits of Florida and from Central America by the 210 km wide Yucatán Channel. Grenada, the southernmost island of the Windward Islands, lies only 145 km away from Trinidad which itself is situated adjacent to the coast of Venezuela.

Furthermore, the close location of the islands one to each other is also of zoogeographical importance. For those times, in which the land connection on the Central American land brigide had been interrupted (WEYL 1970), it has been considered probable that a migration route existed between North and South America via the Greater and Lesser Antilles. DARLINGTON (1957) calls Cuba the 'port of entry'

69

for immigrants from the north and south. In times, however, when the sea-level was higher than today, some of the Greater Antilles Islands also consisted of several islands. In spite of this, a spreading from island to island, especially of animals capable of flying, is very well imaginable (CARY 1951). For the moment, however, these questions are only of secondary importance in the present investigations.

A feature which the West Indian Islands have in common with the Galápagos Islands is their insular character, which offered the settling species ideal conditions to develop divergently during isolation, as shown by the endemic sphingids

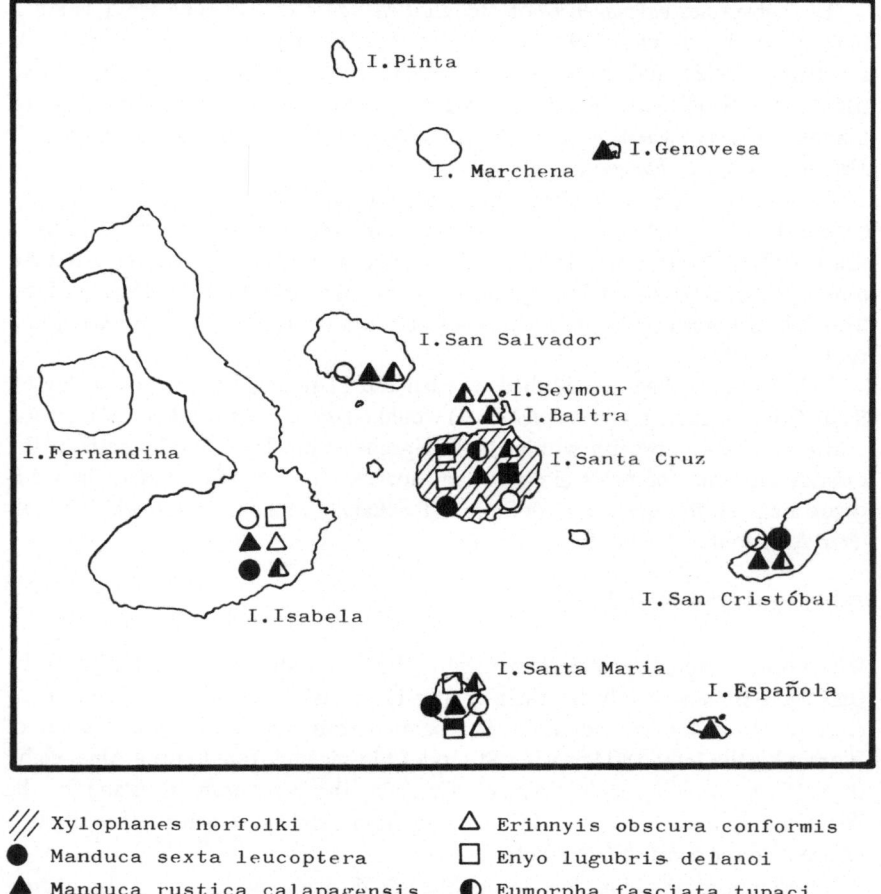

Fig. 10. Distribution of endemic sphingids of the Galápagos Islands.

occurring on the West Indian Islands. The distribution of the sphingids on the Antilles has been dealt with by CARY (1951). Entomologically, the West Indian Islands have been studied, among others, by BATES 1935, CARY 1951, CLENCH 1963, DARLINGTON 1938, FORBES 1930, GEIYSKES 1934, GROTE 1865, GUNDLACH 1881, DE LA TORRE 1960, ZAYAS & ALAYO 1956.

On the basis of the mapped ranges of West Indian sphingids, centres of distribution could be established on Cuba, Jamaica, Hispaniola and the Lesser Antilles.

2. The East Cuban centre

With its 109,216 km², Cuba is by far the largest of the West Indian Islands. This very long-stretching island, with 1,200 km in length and 30-145 km in width, consists, in contrast to almost all the remaining Antilles islands, mainly of hilly lowland reaching heights of 2,000 m only in the Sierra Maestra in the southeast, 1,500 m in the middle south and 700 m in the west. The mapped localities show a correlation with the height level in that they frequently avoid the lowland.

Strays deducted, 59 different sphingids are known from Cuba, of which 45 also occur on one or more of the remaining islands. This means that more than half of the 105 sphingids shown to live on the West Indian Islands inhabit Cuba. This figure does not include 18 species and subspecies which have been found in one or two specimens on the island, but otherwise occur on the mainland, only.

The following 14 sphingids are species and subspecies endemic on Cuba:

1. Cocytius vitrinus
2. Manduca rustica cubana
3. Amplypterus gannascus cubanus
4. Isognatus rimosus rimosus
5. Pachylia syces cubensis
6. Callionima gracilis
7. Callionima ramsdeni
8. Perigonia divisa
9. Aellopos titan cubana
10. Xylophanes gundlachi
11. Xylophanes chiron cubanus
12. Xylophanes tersa cubensis
13. Xylophanes robinsoni
14. Hyles calverleyi

Among the seven endemic monotypic species, *Xylophanes gundlachi* and *Xylophanes robinsoni* are spread all over Cuba (figs. 11, 35, 36, pp. 73, 143, 144).

The ranges of *Cocytius vitrinus* and *Callionima ramsdeni*, however, are strictly confined to the eastern part of the island, thus marking a possible centre of distribution in this area, the existence of which is further supported by the fact that the ranges of other monotypic species like *Enyo boisduvali* and *Perigonia lefeburei*, all of whose exact localities are to be found in the eastern part of the island, also partly cover the area in question. The existence of a centre of distribution in this part of Cuba can be further substantiated by the ranges of subspecifically differentiated sphingids. *Enyo boisduvali* has originally been described from Surinam, it is true, but its occurrence on the continent was later confirmed only once by a specimen found in French Guiana. The 27 specimens for Cuba were recorded from eight different collections.

The concentration of the distribution range of *Perigonia lefeburei* is in East Cuba and West-Hispaniola, i.e. Haiti (fig. 37, p. 145). *Callionima gracilis, Perigonia divisa* and *Hyles calverleyi* could not be evaluated chorologically, since they are hardly documented and possibly even doubtful. (*Hyles calverleyi* described by GROTE in 1865, was neither available to ROTHSCHILD & JORDAN 1903 for their revision, nor present in any of the large sphingid collections examined). Among the polytypic species, the subspecies endemic for Cuba occur all over Cuba, with the exception of *Aellopos titan cubana* and *Xylophanes chiron cubanus* whose main distribution is definitely concentrated in the east part of the island (fig. 11)*.

The ranges of sphingids which, besides Cuba, also occur on the neighbouring islands or the adjacent mainland, have either a share in Hispaniola, like *Manduca afflicta, Erinnyis guttularis* (fig. 38, p. 146) or *Perigonia lefeburei*, or they show relations to Florida and the Bahamas, like *Manduca brontes cubensis* or *Madoryx pseudothyreus* (fig. 39, p. 147). *Xylophanes irrorata* occurs, apart from Cuba, only on the Bahamas and could possibly have differentiated there, as could be inferred from the localitydata (fig. 12).

On the Bahamas, *Xylophanes suana* occurs as an endemic monotypic species, *Manduca afflicta bahamensis*, whose closest relatives occur on Cuba and Hispaniola, inhabits the Bahamas as a genuine subspecies. *Perigonia lusca bahamensis* which has been described by CLARK (1917) as an individual insular race from the Bahamas, has been considered synonymous to the nominate form by HODGES (1971). Also, there are species like *Protambulyx carteri*, which have a common occurrence on the Bahamas and in South Florida. I would not yet consider the evidence supplied on the basis of sphingid ranges sufficient to confirm a separate centre of distribution on the Bahamas.

On Cuba as compared with Hispaniola and Jamaica, where each island as a whole has to be looked upon as a centre of distribution, the distribution centre is restricted to the eastern part of the island.

3. The Hispaniola centre

Hispaniola which, in comparison with Cuba, shows a more pronounced orographic pattern, certainly consisted of several centres of distribution in the past. Due to the locomotion capacity and the ecological valency of the species in question, however, as well as due to the anthropogenic restructuring of the natural land-

---

* Inaccurate locality data could not be included in the illustration, yet 50% of them bear the mention 'East Cuba'.

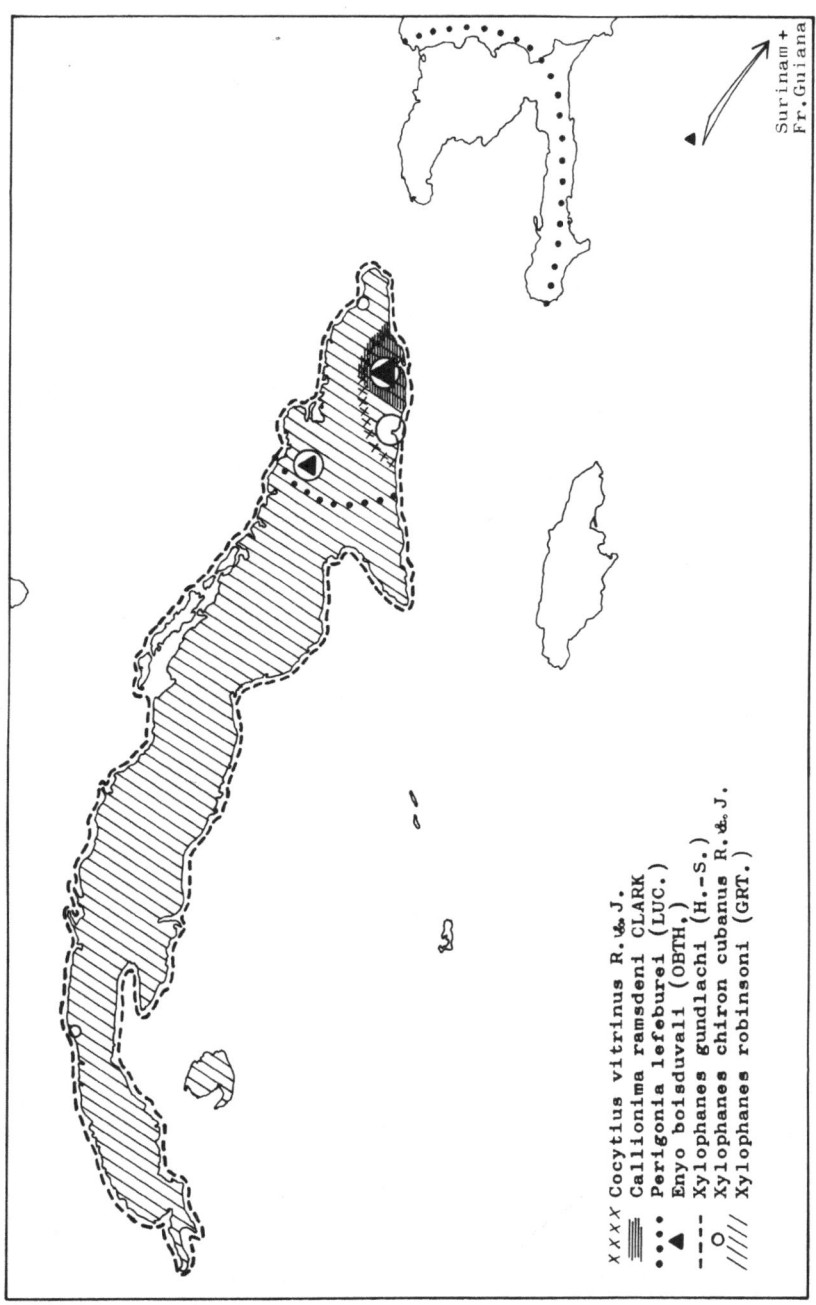

Fig. 11. The East Cuban centre.

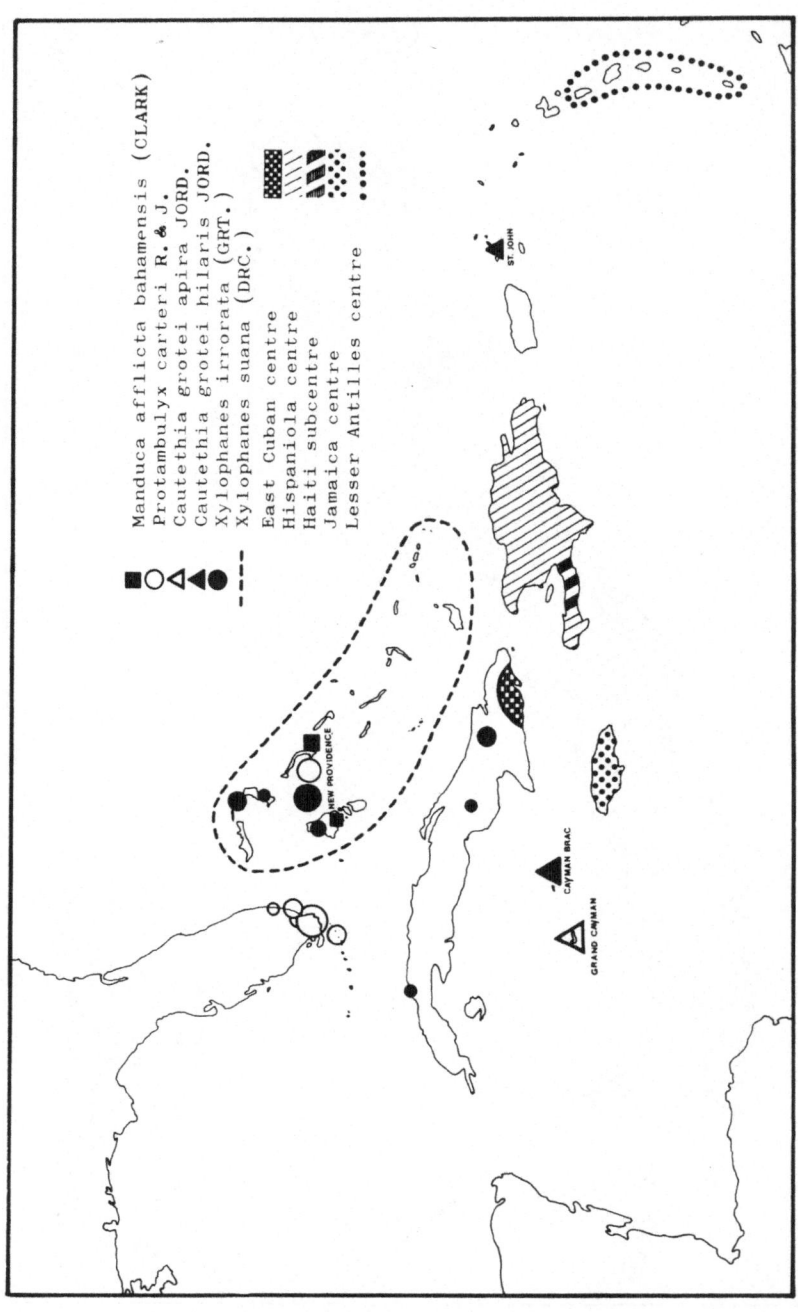

Fig. 12. The centres of the West Indian Islands.

scape, the isolation conditions prevailing in the interior of the island have been weakened. According to 'Meyers Kontinente and Meere' (1969), the natural vegetation is being burned down in the Dominican Republic for cultivation purposes, which contributed greatly to the spreading of pines.

On Haiti, too, cultivation measures are said to have caused such a bad decimation of the original vegetation, except that of the dry areas, that even on steep slopes the wood cover has been destroyed to such an extent that sheet erosion is a widely spread phenomenon in the affected areas. In the southern mountain massifs, pinewoods with *Pinus occidentalis* extend over the hills. This is an element of the Holarctic flora which spread on the Antilles, however only on Cuba and Hispaniola, during the glacial periods of the Pleistocene*.

Excluding strays, 47 specifically and subspecifically differentiated sphingids occur on Hispaniola. The following sphingids are endemics of the island:

1. Manduca caribbea
2. Manduca afflicta johanni
3. Manduca rustica dominica
4. Manduca brontes haitiensis
5. Isognatus rimosus molitor
6. Perigonia manni
7. Perigonia glaucescens
8. Xylophanes rhodocera

Sphinx tricolor, an endemic of the Dominican Republic part of Hispaniola judging by the locality data I perused, is said by CARY (1951) to occur also on the Lesser Antilles. The range of *Eumorpha strenua* includes Hispaniola and East Cuba.

The connection between Haiti and the East Cuban centre has already been demonstrated by the range of *Perigonia lefeburei*. *Pachylia syces insularis* occurs on Hispaniola and Jamaica. On the bais of the ranges of the monotypic species *Perigonia manni* and *Xylophanes rhodocera* as well as the endemic subspecies *Manduca rustica dominica, Manduca brontes haitiensis* and *Isognathus rimosus molitor* which also spread all over Hispaniola, a centre of distribution covering the whole island can be established (fig. 13).

Manduca caribbea was only found in the south of Haiti, *Manduca afflicta johanni* on Haiti only.

The majority of the localities of *Perigonia glaucescens* and *Sphinx tricolor*, on the other hand, lie in the Dominican Republic; *Sphinx tricolor* is said to have also been found on the Lesser Antilles.

Since the sphingid ranges do not allow any final conclusions, I would like to speak, besides the Hispaniola centre extending over the whole island, of a Haiti-subcentre only, which is probably located in the south of Haiti, as indicated by the range of *Manduca caribbea*.

* from 'Meyers Kontinente und Meere' (1969)

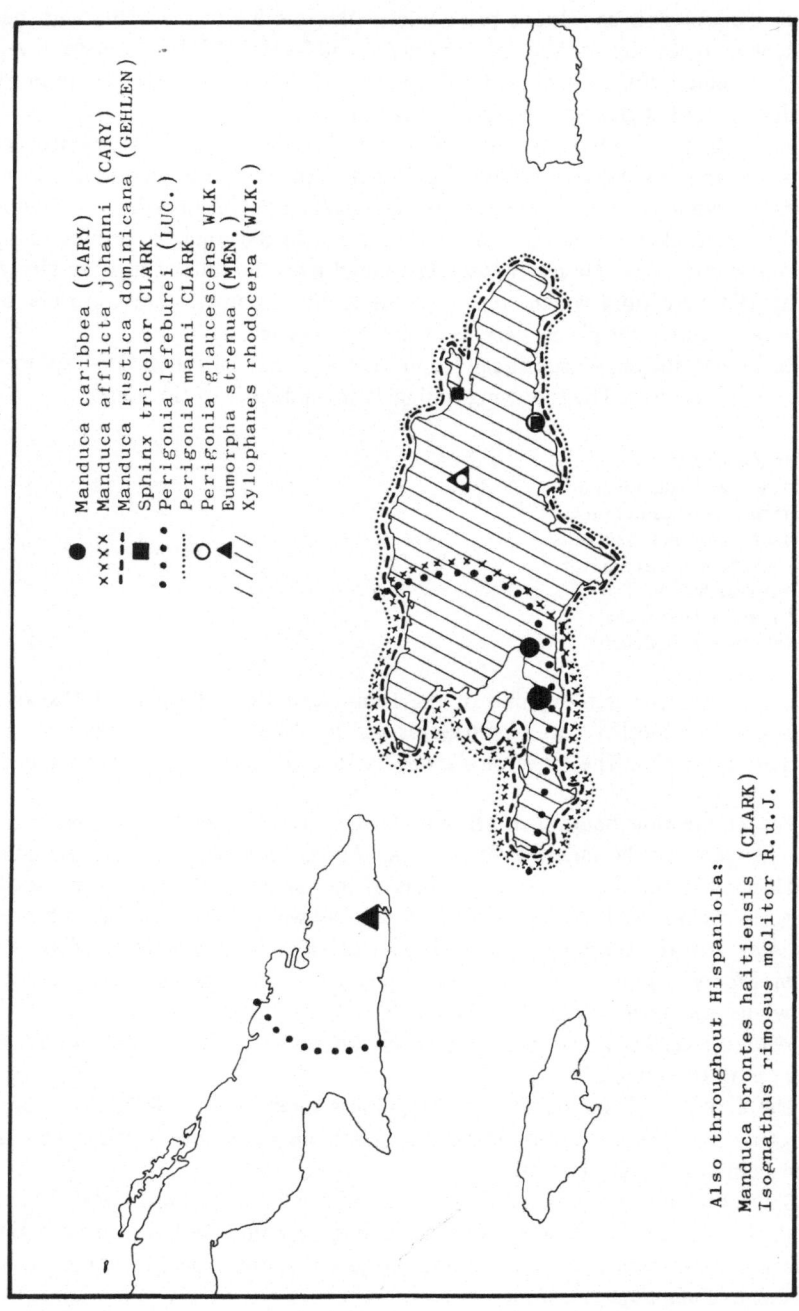

Fig. 13. The Hispaniola centre.

4. The Jamaica centre

With a surface of 11,424 km², Jamaica is much smaller than the two-states-island of Hispaniola (76,484 km²) discussed in the preceding chapter. Compared with Cuba and Hispaniola, the number of sphingids found on Jamaica is amazingly high. It amounts to a total of 43 species and subspecies. This figure does not include 7 taxa which have to be considered as strays. *Xylophanes jamaicensis* seems doubtful. The relatively high number of species as compared with the size of the island could partly be explained by the more southern position of the area which might also have had a favourable effect under palaeoclimatic conditions. Another reason could be the isolation period. CARY (1951) says: 'Jamaica is probably one of the oldest of the Greater Antilles' and 'it has been isolated for so many geological years that its sphingid population has a definitely insular character'.

This finds expression in the number of endemics among the sphingids known from Jamaica which even include one endemic genus.

The following specifically and subspecifically differentiated sphingids are to be considered endemics of Jamaica:

1. Manduca brontes brontes
2. Amplypterus gannascus jamaicensis
3. Isognathus rimosus jamaicensis
4. Erinnyis obscura jamaicensis
5. Callionima elainae
6. Perigonia jamaicensis
7. Enyo lugubris latipennis
8. Eumorpha vitis hesperidum
9. Himantoides undata
10. Himantoides perkinsae
11. Xylophanes chiron chiron

Himantoides is an endemic genus of Jamaica. Its existence could be taken as evidence of the high isolation period of Jamaica. *Himantoides undata* is well documented, whereas *Himantoides perkinsae* is known to me only from 4 specimens.

Besides the two *Himantoides* species, *Callionima elainae* and *Perigonia jamaicensis* are monotypic endemic species. Neither they nor the ranges of the endemic subspecies point to the position of a centre of distribution in any specific part of the island. Nevertheless, Jamaica as a whole is well documented as a distribution centre by the ranges of the sphingids occurring there (fig. 14).

The range of *Pachylia syces insularis* extends to Hispaniola, with its centre on Jamaica. *Erinnyis domingonis pallescens* is known from Jamaica and Mexico, *Enyo lugubris latipennis* also from Grand Cayman and, as a stray, from Cuba.

Sphingids recorded from Puerto Rico

Contrary to Jamaica, Puerto Rico is very poor in sphingids. Only 27 specimens have been recorded. Moreover these belong, for the most part, to widely spread

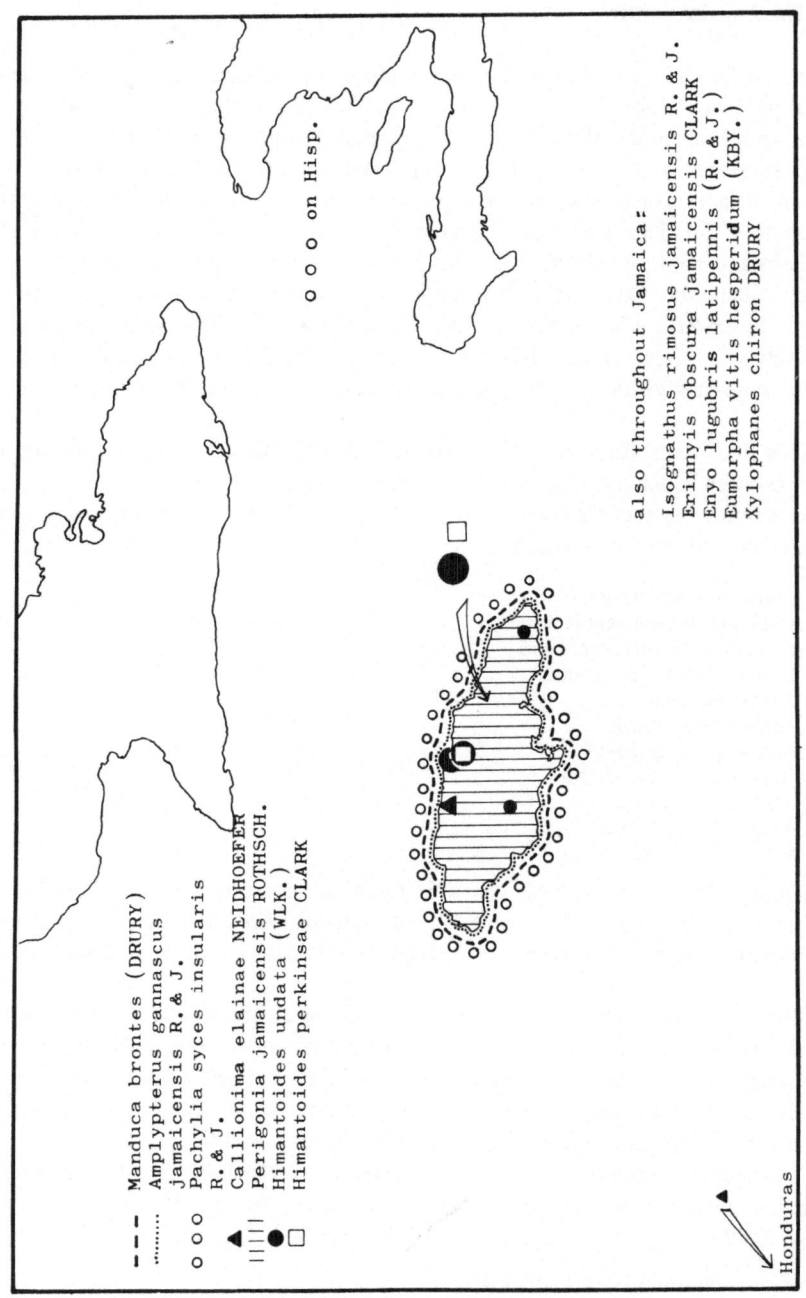

Fig. 14. The Jamaica centre.

species. Only two subspecies, *Manduca brontes smythi* and *Isognathus rimosus wolcotti*, of which the former only is well documented, are endemics of Puerto Rico (fig. 15). *Protambulyx strigilis portoricensis* has been described by CLARK (1931) on the basis of one specimen from Puerto Rico (Type 4467,♀, CM). There are two further specimens in the USNM determined as *Protambulyx strigilis portoricensis*, one from Cuba, the other from Haiti. The evidence does not suffice to postulate a centre of distribution for Puerto Rico.

5. The Lesser Antilles centre

Excluding 2 strays, I found 38 different sphingids confirmed on the Lesser Antilles.

Perigonia thayeri is specifically differentiated on St. Vincent. *Manduca rustica harterti, Eumorpha vitis fuscata, Cautethia noctuiformis bredini* (from Antigua and Barbuda), *Xylophanes chiron martiniquensis* and *Xylophanes chiron lucianus* are endemic subspecies of the Lesser Antilles (fig. 16). The number of endemics given by CARY (1951) for the Lesser Antilles (7 species and 7 subspecies) is higher than the one I found, for two reasons: first, the author includes the Bahamas in the Lesser Antilles, and secondly, her list includes sphingids as endemics which also occur on other West Indian Islands or, even predominantly, in Florida, such as *Cautethia grotei*. In addition, CARY (1951) considered *Manduca sexta lucia* and *Perigonia lusca major* as subspecies. KERNBACH (1964b) cast doubt on the existence of *Manduca sexta lucia* on the basis of genital studies. *Perigonia lusca major* is not a valid subspecies either. *Erinnyis stheno* has been considered by HODGES (1971) as a synonym of *Erinnyis obscura*, which seems correct in view of the mapped distribution. (The specimens of *Erinnyis stheno* documented from St. Vincent, Barbados, Hispaniola and the Bahamas besides *Erinnyis obscura* have been mapped as *Erinnyis obscura stheno*.)

I would like to define the borders of a centre of distribution on the Lesser Antilles on the range of *Eumorpha vitis fuscata* (fig. 40, p. 148), even though this limitation fails to allow for the place of differentiation of some individual taxa.

Trinidad and Tobago Island have not been included in the Antilles, since, in doing so, I would have had to consider another 44 Neotropical sphingids which also occur on the mainland.

The Central American centres

Islands are clearly defined as given spatial units by their natural borders which frequently also correspond at least to the recent outlines of the ranges of the taxa inhabiting them. In contrast to this, comparable areas on the mainland with which a correlation of as many organisms as possible should be proved to exist, must first be worked out.

The way to this end leads over the plotting of the distribution ranges of species with narrow ranges on a map of the region, one above the other.

In comparison with the West Indian Islands, the Central American-Mexican

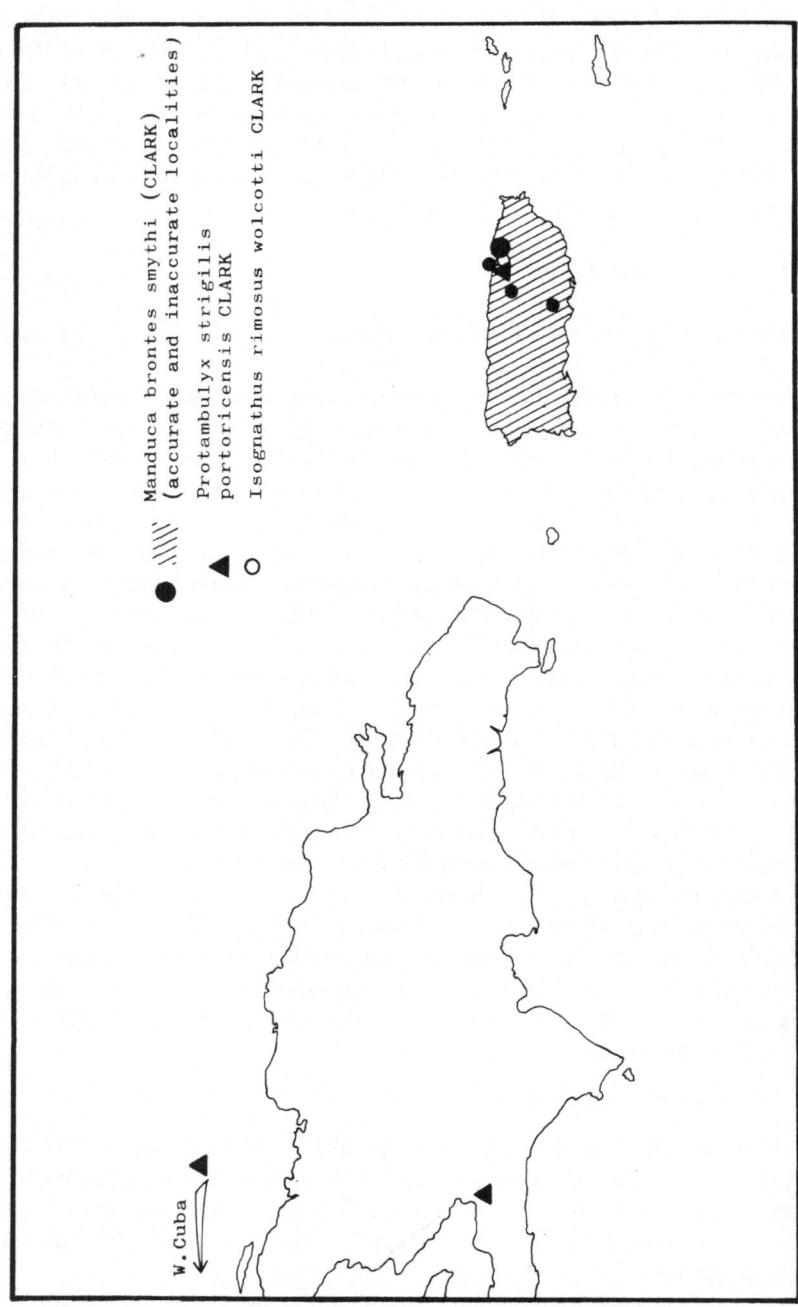

Fig. 15. Sphingids recorded from Puerto Rico.

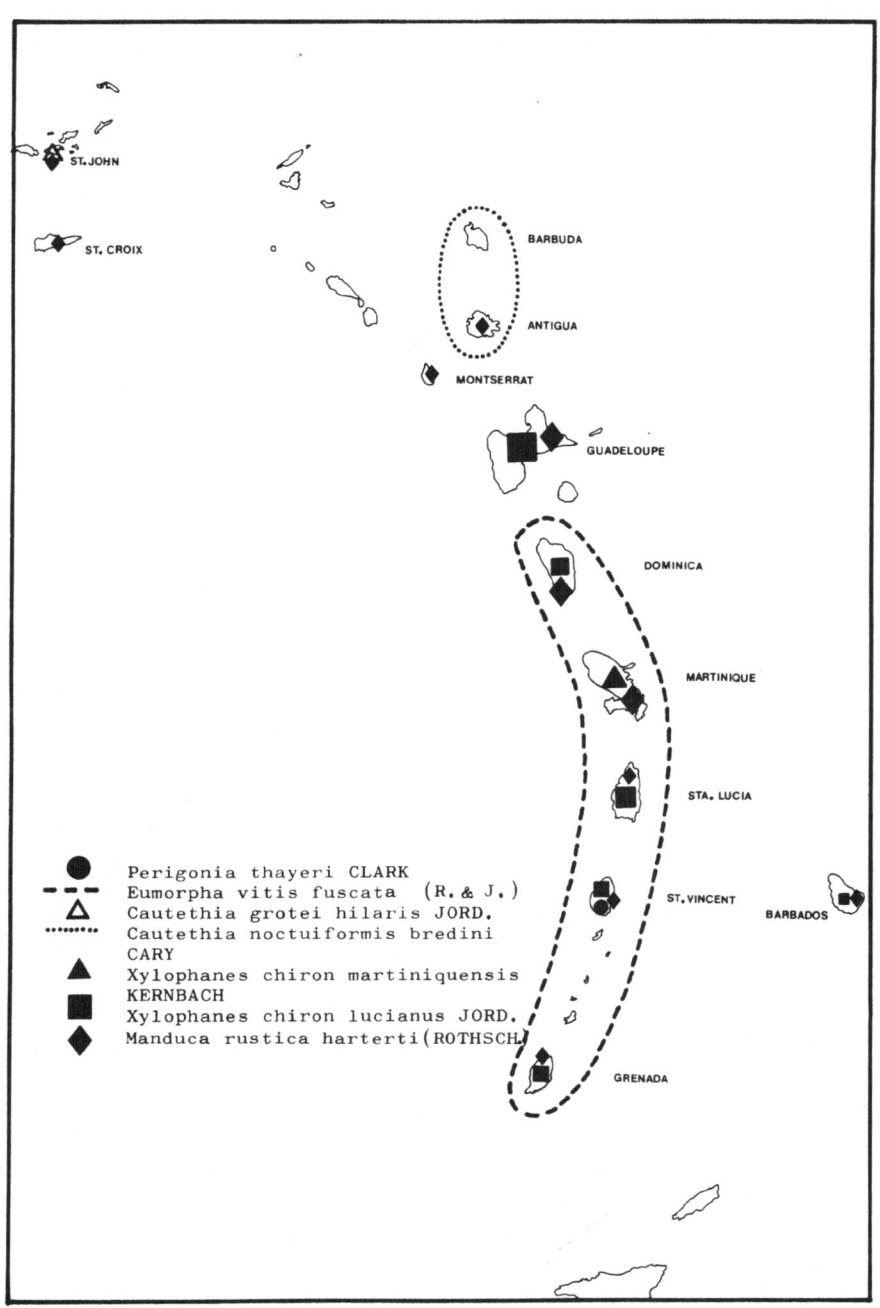

Legend within figure:

● Perigonia thayeri CLARK
– – Eumorpha vitis fuscata (R. & J.)
△ Cautethia grotei hilaris JORD.
•••••• Cautethia noctuiformis bredini CARY
▲ Xylophanes chiron martiniquensis KERNBACH
■ Xylophanes chiron lucianus JORD.
◆ Manduca rustica harterti (ROTHSCH.)

Map labels: ST. JOHN, ST. CROIX, BARBUDA, ANTIGUA, MONTSERRAT, GUADELOUPE, DOMINICA, MARTINIQUE, STA. LUCIA, ST. VINCENT, BARBADOS, GRENADA

Fig. 16. The Lesser Antilles centre.

region is especially distinguished by a greater richness in species. Depending on their smaller land-masses and, hence, their restricted ecological species diversity, islands on the other hand, are characterized by a relative poorness of species compared to corresponding areas on the neighbouring mainland (DE LATTIN 1967).

In Mexico and Central America, 230 specifically and subspecifically differentiated sphingids were recorded. This number, which is very high even in comparison with South America, cannot be explained by reasons of recent ecology alone. It also throws light on the functions this area had to fulfil as a centre where faunas were preserved and have speciated in the past.

Both the location and the orography of the area might have been important factors in this regard. In Mexico, which belongs to the Nearctic region as far down as to the Balsas valley excluding the coastal lowland, and on the Central American land bridge, the mountain ranges of western North America extend over a length of 4,300 km enclosing, in Mexico, a high plateau between the Sierra Madre Occidental and the Sierra Madre Oriental with heights of more than 5,500 m.

As can be seen from the mapped ranges of the sphingids occurring in this area, the localities are mainly found in arboreal regions referred to as tropical lowland rain forests, tropical mountain rain forests and partly also sclerophyll forests on respective vegetation maps (cf. SCHMITHÜSEN 1969).

The localities are less frequent in areas for which open formations are mentioned. Collecting with the light trap, however, might sometimes create the wrong impression that a large number of species are ecologically adapted mainly to the open terrain, whereas, in reality, this is due to the obstructions to vision in forests. The best results in my experience are reached on elevated locations situated close to the forest. CARCASSON (1968a) also made similar observations in Africa; he says: 'When the lamp is placed inside the forest, results are invariably disappointing, probably due to poor visibility and to the fact that most moths fly above the canopy'.

In defining the border between the area under investigation and the Nearctic region, several authors who studied the Mexican-Central American region have been mentioned (cf. also BARRERA 1962, GOLDMAN & MOORE 1946, HOFF-MANN 1942, KRAUS 1955 and 1960, MAYR 1964, MOOSER 1940 and 1942, RYAN 1963, SAVAGE 1966, SMITH 1940, WELLING 1966, WEYL 1956 and WOODRING 1966).

The attempts especially by SMITH (1940) and by GOLDMAN & MOORE (1946) at dividing Mexico into biotic provinces illustrate the methodical difference from the concept of dispersal centres. The latter tries to illuminate the correlations expressing themselves in a common nuclear area of different taxa rather than accepting recent factors as limiting the range borders.

Starting from the Mexican arboreal centre (cf. fig. 17) which DE LATTIN (1957) considered to be the southernmost centre of dispersal of the Nearctic region, the mapping of the ranges of Neotropical sphingids led to the establish-

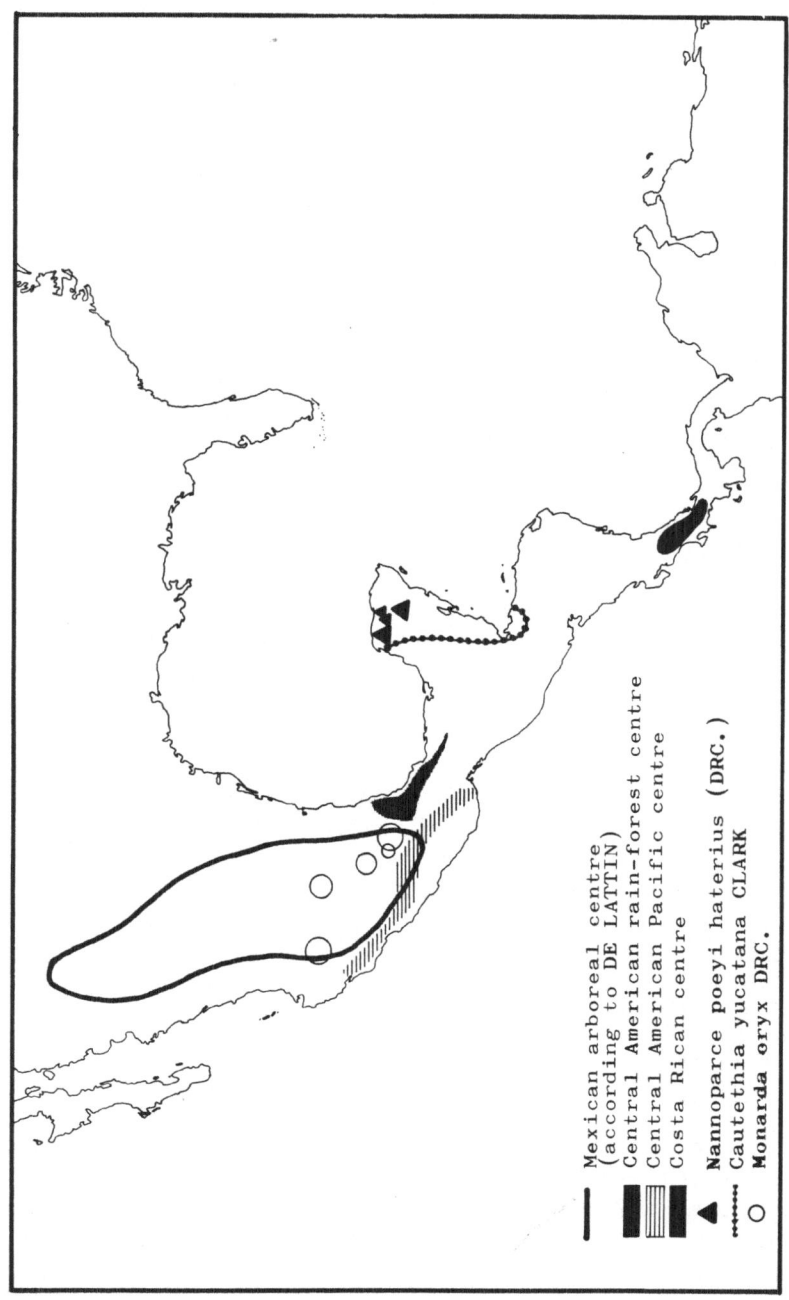

Fig. 17. Dispersal centres in Central America.

ment and confirmation, respectively of another 13 centres of distribution in Central and South America which will be discussed in the following.

6. The Central American rain-forest centre

The Central American rain-forest centre is situated on the Atlantic coast of Mexico in the State of Vera Cruz, thus lying in the so-called 'Biotic Province of Vera Cruz' (cf. GOLDMAN & MOORE 1946). It is identical with the homonymous centre established by MÜLLER (1970) on the basis of vertebrate ranges. The main centre of distribution, which might partly also represent a place of especially thorough collecting, lies between Jalapa, Orizaba and Vera Cruz. As seen from the altitudes of localities above sea-level, the centre includes both the lowland rain forest and the montane forest zone up to elevations of 1,500 m. The centre can be defined by the overlap of the ranges of *Sphinx leucophaeata*, *Nyceryx muelleri* and *Xylophanes damocrita*, which are monotypic species and have to be regarded as endemics of this area due to their small ranges of distribution (fig. 18). *Xylophanes juanita* occurs also beyond the isthmus of Tehuantepec, in Guatemala, British Honduras and El Salvador. *Erinnyis yucatana* has its unequivocal main distribution in the Central American rain-forest centre (fig. 41, p. 149). This species however also penetrated into the Balsas valley as far as the west coast and occurs also on the peninsula of Yucatán. *Manduca dilucida* has a similar distribution. The ranges of *Callionima falcifera* and *Sphinx adumbrata*, too, have a share in the Balsas valley. *Mandura azteka* from Taumalipas, *Ceratomia hoffmanni*, *Xylophanes hannemanni* and *Xylophanes muelleri* from Vera Cruz, *Sphinx pseudostigmatica*, *Sphinx chisoya* and *Nannoparce balsa* from the area of the Balsas valley as well as *Manduca crocala* and *Sphinx praelonga* from Guatemala and Honduras were recorded only in a small number of specimens.

Furthermore, the following subspecies have to be regarded as faunal elements of the Central American rain-forest centre:

1. Manduca hannibal mayeri
2. Manduca lefeburei bossardi
3. Manduca florestan ishkal
4. Amplypterus donysa donysa
5. Darapsa myron mexicana
6. Xylophanes thyelia salvini

The mentioned subspecies underwent subspecific differentiation in the Central American rain-forest centre. In addition to wide-spread species, *Amplypterus globifer*, *Madoryx pluto dentatus* and *Xylophanes amadis cyrene* have a share in the described centre. They occur disjunctly also in the Costa Rican centre, *Stolidoptera tachasara* and *Xylophanes turbata* in addition to this also in the Venezuelan centre (fig. 42, p. 150). *Xylophanes neptolemus* (fig. 43, p. 151) ocuurs, more or less uninterruptedly, mainly in the montane rain-forest biomes of Central America and the Venezuelan coastal mountain ranges.

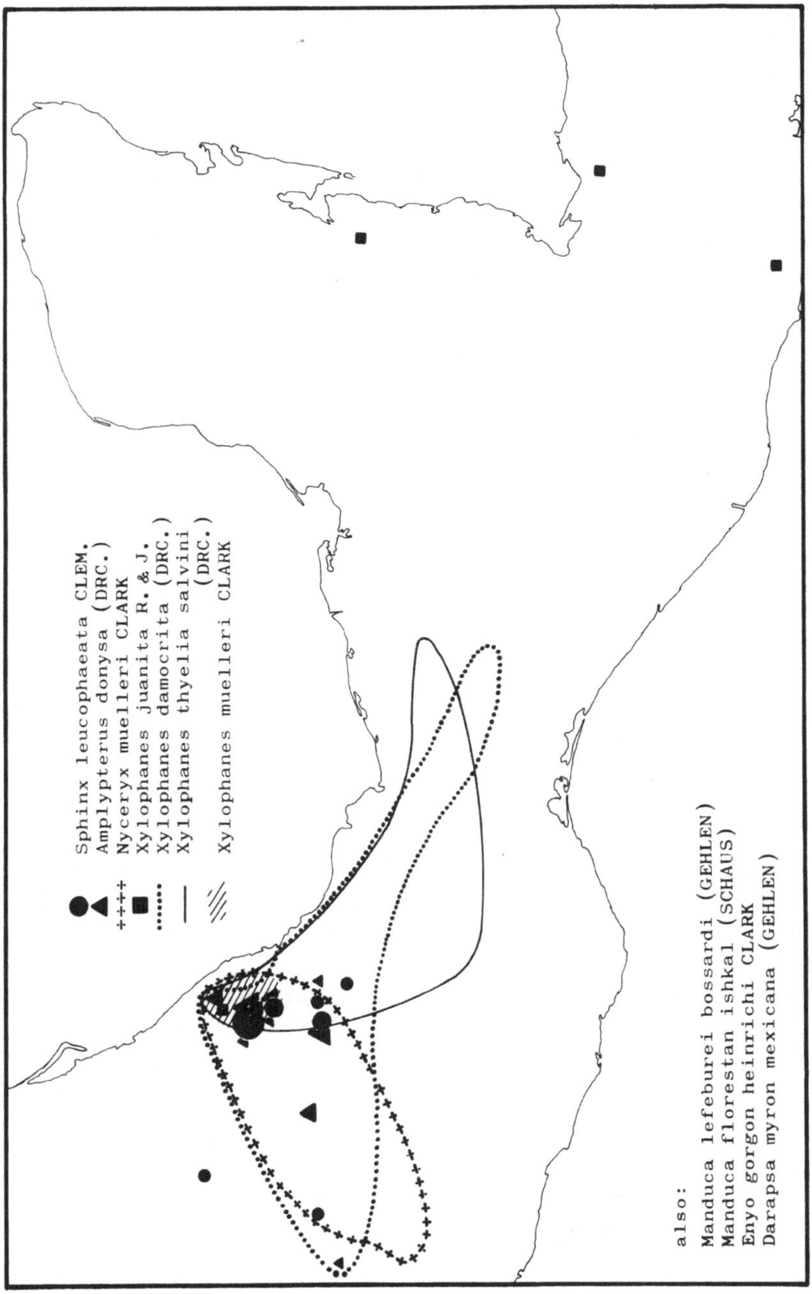
Sphinx leucophaeata CLEM.
Amplypterus donysa (DRC.)
Nyceryx muelleri CLARK
Xylophanes juanita R. & J.
Xylophanes damocrita (DRC.)
Xylophanes thyelia salvini (DRC.)
Xylophanes muelleri CLARK
also:
Manduca lefeburei bossardi (GEHLEN)
Manduca florestan ishkal (SCHAUS)
Enyo gorgon heinrichi CLARK
Darapsa myron mexicana (GEHLEN)

Fig. 18. Faunal elements of the Central American rain-forest centre.

7. The Central American Pacific centre

On the Pacific side of southern Mexico, the ranges of sphingids indicate a centre of distribution which partly extends over the southwestern slopes of the Sierra Madre del Sur and partly into the Balsas valley and beyond it northwards. Along the Rio Balsas some species, such as *Ceratomia igualana*, penetrate far into the east (fig. 19). The range of *Manduca barnesi*, on the other hand, extends on the Pacific side from Guatemala up to Sinaloa.

The centre established on the basis of sphingid ranges is coincident with the Central American Pacific centre postulated by MÜLLER (1970). It is situated within the biotic province of Nayarit-Guerrero of GOLDMAN & MOORE (1946). The centre can, however, not be regarded as completely forest-free in view of the ecological valency of the sphingids occurring there. The Central American Pacific centre can be defined on the coinciding ranges of *Manduca barnesi*, *Dolbogene hartwegii*, *Ceratomia igualana* and *Sphinx balsae*. *Manduca kuschei*, *Manduca holcombi*, *Sphinx biolleyi*, *Gurelca muelleri* and *Xylophanes katharinae* have also been described as occurring in this area, but they are poorly documented.

The following subspecies have to be added as faunal elements of this centre:

1. Manduca occulta pacifica
2. Manduca sesquiplex sesquiplex
3. Madoryx bubastus butleri
4. Pachylia syces septendrionalis
5. Proserpinus terlooii mooseri

It is possible that *Pachysphinx modesta regalis*, too, has differentiated subspecifically in this area.

An overlap with the Nearctic Mexican centre in the area of the Balsas valley which might be conspicuous at first glance, does not stand closer investigation, since the localities of the sphingid taxa found here lie 1,000 m higher on the average. Among them, there is at least one endemic monotypic genus, i.e. *Monarda oryx* — the placing of the likewise monotypic *Trogolegnum pseudambulyx* is problematic (cf. discussion).

The following sphingid species and subspecies have to be considered as faunal elements of the Mexican arboreal centre:

1. Manduca leucophila
2. Sphinx lugens
3. Sphinx geminus
4. Sphinx separata melaena
5. Sphinx istar
6. Smerinthus saliceti
7. Monarda oryx
8. Eumorpha elisa
9. Xylophanes falco

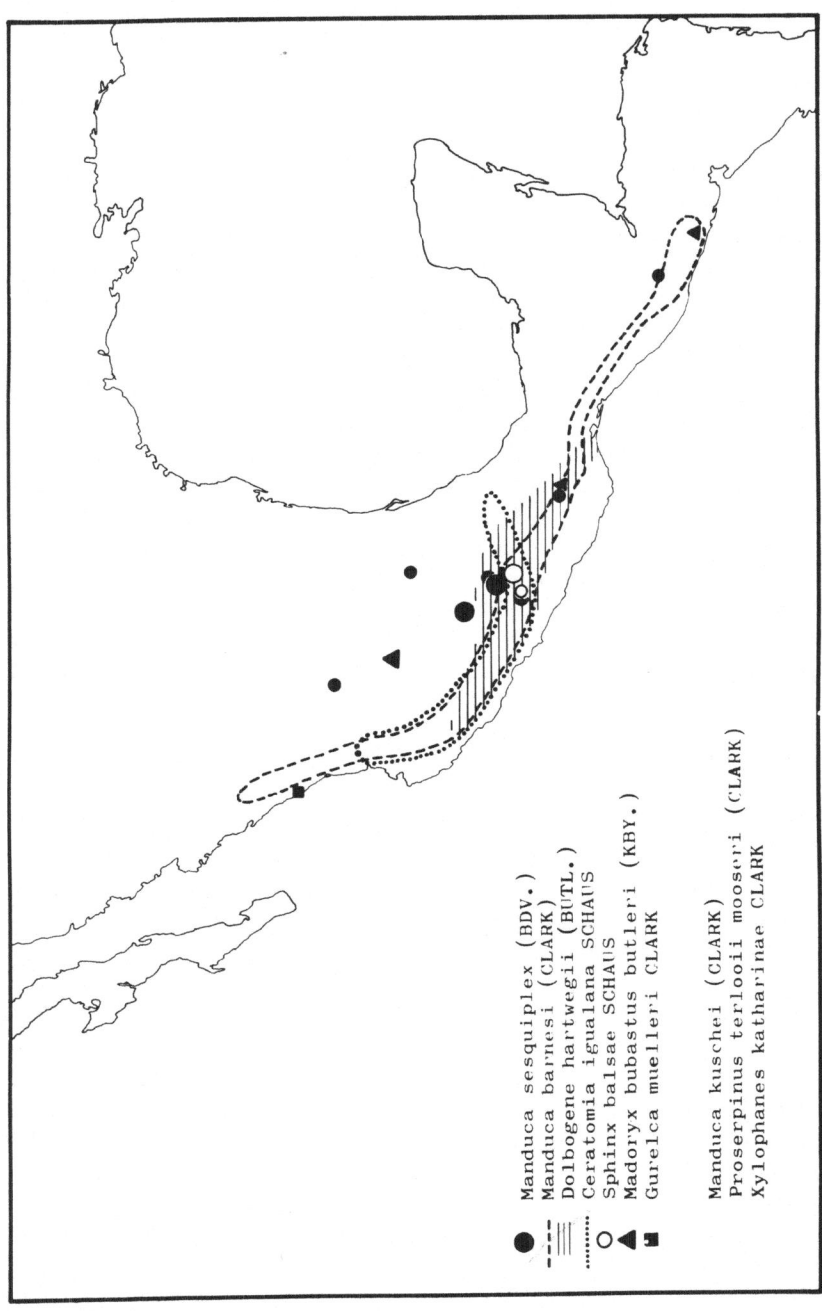

Fig. 19. Faunal elements of the Central American Pacific centre.

Some scattered localities lie outside the main centre of distribution in the Sierra Madre del Sur and in Guatemala.

8. The Yucatán centre

Evidence of sphingids is weak on the peninsula of Yucatán. However, the distribution of *Nannoparce poeyi haterius* (the nominate form occurs on Cuba, Jamaica and Hispaniola) and of *Cautethia yucatana* (fig. 17) can be understood as further supporting evidence of the Yucatán centre postulated by MÜLLER (1970) on the basis of vertebrate ranges. It is conspicuous that many wide-spread species, such as *Agrius cingulatus, Cocytius duponchel* and *Xylophanes tersa* either do not occur on Yucatán at all or only very sparsely. The reason for this phenonemon could be seen in a pronounced anthropogenic restructuring of the original landscape, since, according to WELLING (1966), all of Yucatán had been covered with wood in previous times. The natural vegetation suffered great damage as early as during the Maya period. According to WELLING it is assumed that, during the classic phase of maturity of the Mayan civilization 800 to 2000 years ago, 2 million people have been living on the 200,000 square kilometre large peninsula.

The butterfly fauna, on the other hand, seems to have been affected less by the destruction of the woods, since WELLING mentions 720 different species and assumes the existence of another 150. With regard to those butterflies, however, which like most of the sphingids, prefer shady biotopes, he says that they are hard to collect and retreat to places where there are still 'some trees left'.

Besides the 2 sphingid representatives with small ranges occurring on Yucatán, 20 other specifically and subspecifically differentiated sphingids were recorded (10 others on the basis of one specimen only). The sphingids found in British Honduras are not included in this figure. If one included British Honduras in the faunistic unit of the Yucatán-peninsula, another 54 species and subspecies would have to be considered. Apart from wide-spread polycentric species the sphingids collected in British Honduras show either correlations to the Central American rain-forest centre or to the Costa Rican centre.

9. The Costa Rican centre

The last of the Central American centres which can be established on the basis of the ranges of sphingids, is the Costa Rican centre, which extends from Costa Rica to Panama (fig. 20). Its position is identical to that of the homonymous rain-forest centre established by MÜLLER (1970). This centre clearly shows an abundant occurrence of sphingids with small ranges, a fact, which corresponds to similar observations made on other animal groups like fishes and amphibians (MÜLLER 1970, 1973).

The following species and subspecies can be regarded as faunal elements of the Costa Rican centre:

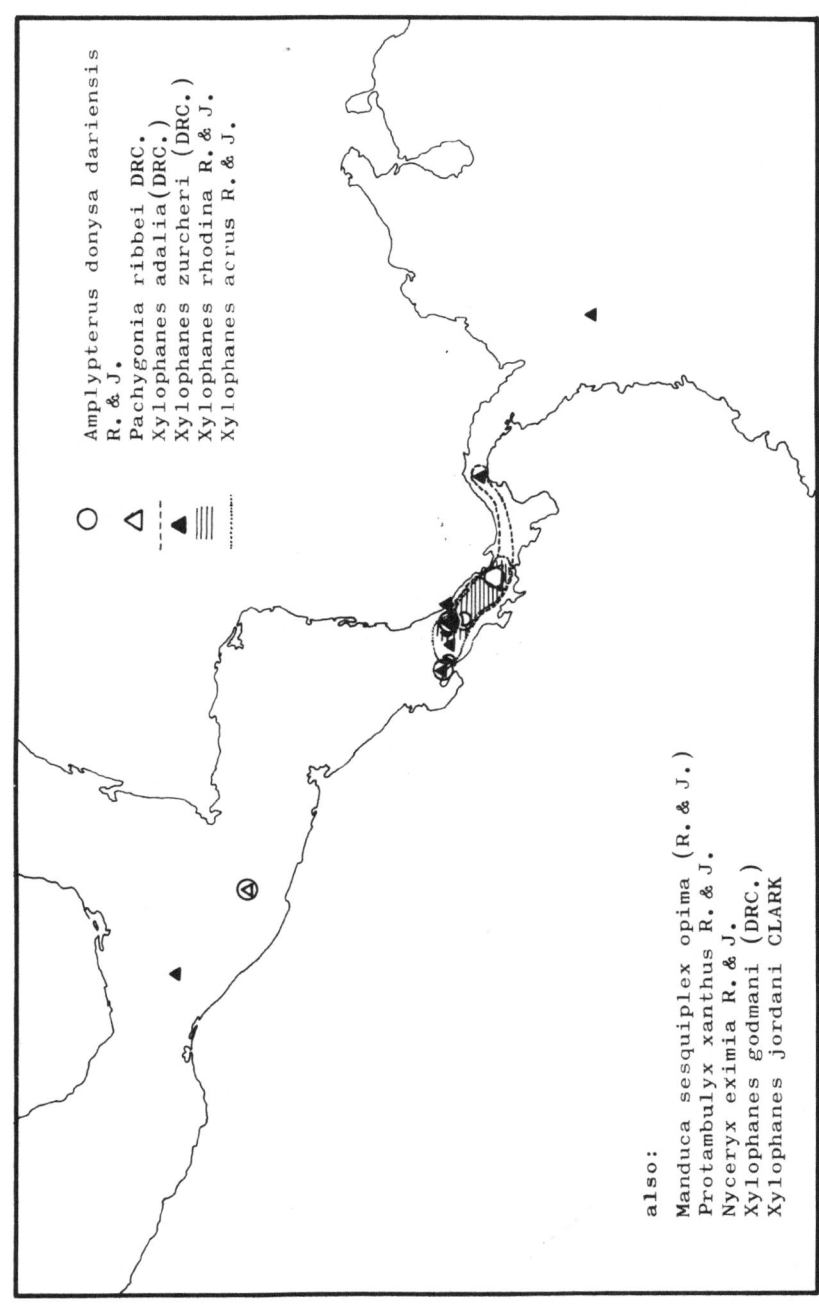

Amplypterus donysa dariensis R. & J.
Pachygonia ribbei DRC.
Xylophanes adalia (DRC.)
Xylophanes zurcheri (DRC.)
Xylophanes rhodina R. & J.
Xylophanes acrus R. & J.

also:
Manduca sesquiplex opima (R. & J.)
Protambulyx xanthus R. & J.
Nyceryx eximia R. & J.
Xylophanes godmani (DRC.)
Xylophanes jordani CLARK

Fig. 20. Faunal elements of the Costa Rican centre.

89

1. Manduca lucetius nubila
2. Manduca sesquiplex opima
3. Protambulyx xanthus xanthus
4. Amplypterus donysa dariensis
5. Pachygonia ribbei
6. Nyceryx eximia
7. Xylophanes adalia
8. Xylophanes zurcheri
9. Xylophanes rhodina
10. Xylophanes godmani
11. Xylophanes acrus
12. Xylophanes belti

Xylophanes jordani has been recorded on the basis of three specimens from Costa Rica and one from Bolivia only.

While the mapped ranges of the monotypic species indicate the exact location of the distribution centre north and south of the Cordillera de Talamanca (MÜLLER 1970, 1973) through their common nuclear area (REINIG 1950), the existing subspecies show that this centre also represents a centre of speciation. The nominate form of *Amplypterus donysa dariensis* (fig. 44, p. 152) is a faunal element of the Central American rain-forest centre, that of *Manduca sesquiplex opima* a faunal element of the Central American Pacific centre. *Manduca lucetius nubila* has its closest relatives in the Andes and in the Serra do Mar as well as in Uruguay, whereas *Protambulyx xanthus* has relatives in the Serra do Mar.

A total of 130 sphingid species and subspecies belong to the Costa Rican centre, another 20 must be regarded as strays. Many Neotropical taxa have their northernmost occurrence in this area. An interesting phenomenon is the disjunct distribution of some species in the Costa Rican centre on the one hand and the Central American rain-forest centre and, in some cases, the Venezuelan centre, on the other.

The richness in taxa seems to be understandable from two factors: first, from the role this area plays as a bridge between the south and north continents, and secondly from the pronounced orographic and climatic differentiation of this area, which finds expression in a great variety of vegetation types and plant societies with great richness in species (cf. Meyers Kontinente und Meere, 1969). The majority of localities of occurrence of sphingids lie in altitudes of approximately 1,000 metres, i.e. a region for which KNAPP (1965) mentioned montane rain forests with oaks of which he says: 'A large part of the famous abundant orchid flora of Costa Rica appear among the epiphytes of these forests'.*

The particularly abundant occurrence of sphingids in montane forests observed in Central America is also confirmed in South America, where the sphingids

---

* 'Ein großer Teil der berühmten reichen Orchideen-Flora von Costa Rica erscheint unter den Epiphyten dieser Wälder'.

are particularly well represented in the montane and submontane forests of the Andes and the Serra do Mar.

The Andean centres

The Andes extend over a length of 7,300 km from South Chile up to Venezuela. They reach heights of more than 6,000 m and have their greatest east-west extension of 800 km in Bolivia.

At the narrowest site in Ecuador, they are only 200 km wide. In Colombia, the Andean chains are widely and deeply intersected by the valleys of the rivers Magdalena, Cauca and Atrato.

The Andes are an insurmountable obstacle to the distribution of numerous species, and they are of influence through the climatic and vegetation zones in their vertical structuring.

A comparison of the species living on the north and south continents has shown that the lower temperatures of higher latitudes have prevented many species from spreading further, and this also repeats in mountains.

Apart from some exceptions, the sphingids avoid dry, forest-free biomes (SEITZ 1931). The distribution of many wide-spread East-Andean species, however (figs. 45, 46, 47, 48, 49, 50, pp. 153-158), has shown that the lowland rain forest, too, can play a decivise role in limiting the ranges of sphingids. This is particularly remarkable, since, according to SCHMITHÜSEN (1968), the forest of the submontane altitudes between approximately 400 and 1,000 m normally does not differ greatly from that of the lower elevations.

MANN (1968), too, writes that 'in the vicinity of the equator the premontane level is occupied by montane forests which are largely composed of basimontane elements and, hence, have a marked habitual similarity to the macrothermic rain forest of the plane'.*

The most frequent Andean localities of sphingids are to be found (fig. 51, p. 159) in heights from 500 to 2,000 m both on the submontane and the montane rain forest level. Towards the south, the localities lie in distinctly lower altitudes which indicates that the level of the vegetation zone, in which sphingids have their privileged occurrence, sinks in proportion to the increasing latitudes (cf. TROLL 1955). Only in northern Ecuador and Colombia, Andean sphingid localities can also be found on the west side of the Andes, apart from localities in the wide river valleys between the Cordilleras.

A total number of 262 specifically and subspecifically differentiated sphingids occur in the Andes (9 of them might only be strays). 96 species and subspecies are exclusively limited to the Andes, another 20 have been recorded on the basis of one specimen only or taken from literature. With a percentage of exactly 50% the Andes have a high share in the sphingid fauna of the Neotropical region. The

---

* '... in Äquatornähe die prämontane Stufe durch Bergwälder eingenommen wird, die zum größten Teil aus basimontanen Elementen aufgebaut sind und damit starke habituelle Ähnlichkeit mit dem makrothermen Regenwald der Ebene aufweisen'.

percentage of 22% of sphingids so far known exclusively from the Andes is likewise high. This oberservation coincides with the findings of authors who worked on different animal groups in the Andes, with the exception of studies which have been restricted to the uplands of the Andes which are faunistically poor for climatic reasons. FITTKAU (1969) refers to the entire Andean-Patagonean area when speaking of a less pronounced species density in comparison to similar geographical regions, but he also mentions the extremely great richness of endemics. ORLOG (1969) referring to the avifauna of South America says: 'Areas rich in endemics are principally found in the Andes ...' An illustration of South American centres of endemism given by him on the basis of the distribution of birds does not, however, include an East Andean centre situated in Bolivia and Peru as it has been found by MÜLLER (1970) through the study and interpretation of vertebrate ranges and as it can be verified by a chorological analysis of the sphingids occurring in this area.

10. The Yungas centre

This centre extends over the east slopes of the Andes and covers a region called 'Yungas' in Bolivia and 'Montaña' in Peru (HUECK 1966). It is occupied by tropical montane rain forests. Climatically, these forests are characterized by the orographic rain which they receive.

The mapping of the ranges of the sphingids which can be regarded as endemics of this region, shows a high degree of conformity with the vegetation mapping of the montane rain forests by SCHMITHÜSEN (1969). Many localities, however, still lie on the submontane level below the 1000-m-line indicating that according to the criteria decisive for the occurrence of sphingids in this area the submontane level is more closely related to the montane rain forest than to the lowland rain forest.

The ranges of the sphingids pertaining to the Yungas centre end at Santa Cruz, where the montane rain forest is interrupted (figs. 21, 52, 53, 54, 55, pp. 93, 160-163). In the north, *Sphinx aurigutta*, *Xylophanes media* and *Xylophanes rhodotus*, for instance, do not extend beyond the Chanchamayo area situated in the Peruvian province of Junin at 11° south latitude. This particular area has always been a popular place of collecting for entomologists, because, as BAUMANN & REISSINGER (1969) say: 'from Lima, it is the nearest virgin forest area of the East Andean slopes'.* Here sphingids were collected by MOSS (cf. MOSS 1912). The ranges of *Protambulyx ockendeni* and *Xylophanes nabuchodonosor* reach to the North Andean Peru, whereas *Manduca andicola*, *Euryglottis guttiventris*, *Perigonia grisea*, *Eupyrrhoglossum corvus* and *Xylophanes docilis* spread as far as southern Ecuador. Some of the mentioned species as well as many species with an overall Andean distribution pattern show a gap of occurrence between 7° and 10° south latitude.

* '... das Lima am nächsten gelegene Urwald-Gebiet des Anden-Ostabhanges ist'

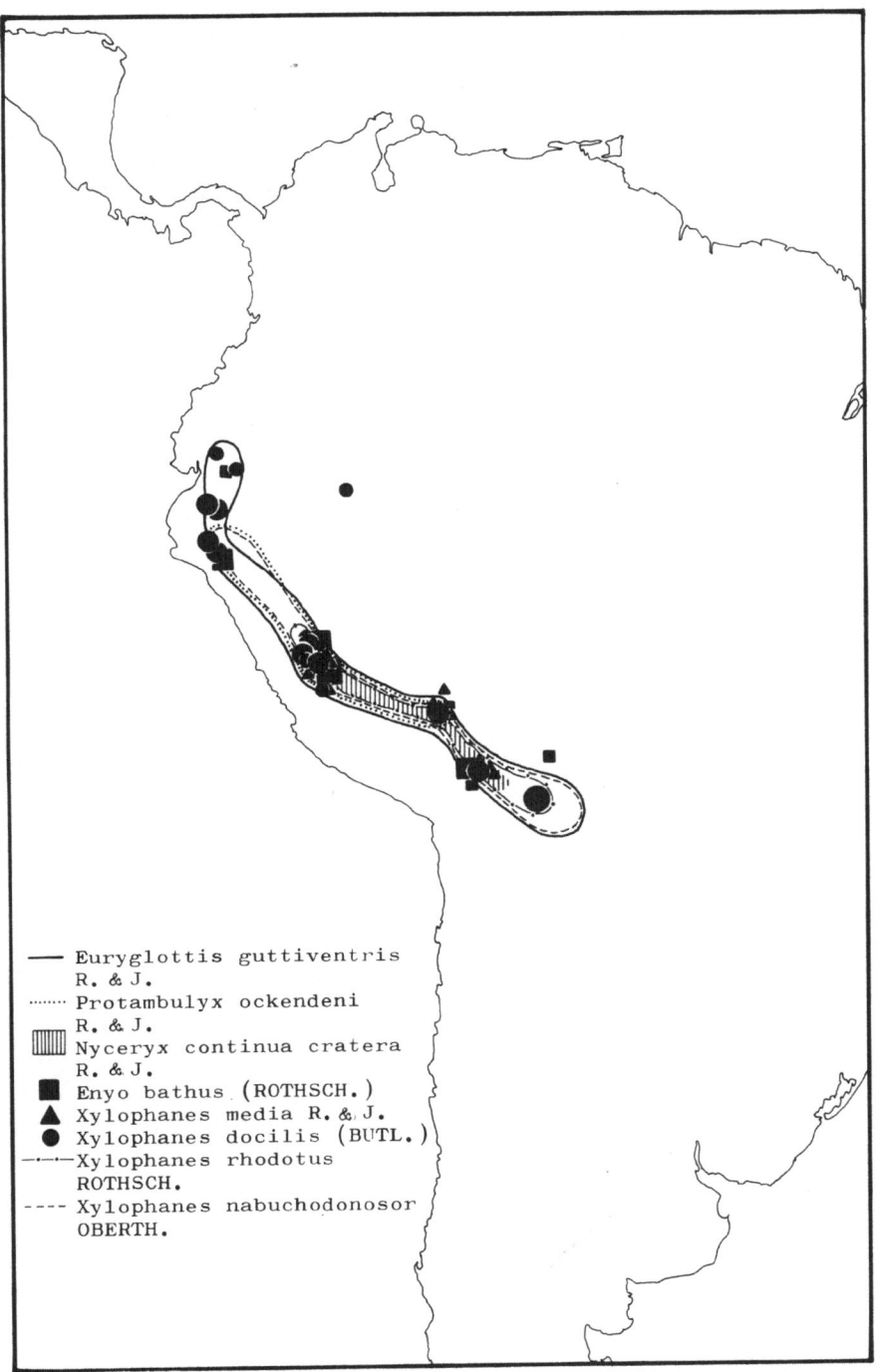

Fig. 21. Faunal elements of the Yungas centre.

This observation could either be explained by less intensive collecting in this area or by reasons of recent ecology or historical reasons. The fact that there are species with a more or less uninterrupted distribution pattern speaks against the former assumption. The assumption of an ecological barrier no longer fully effective recently — the ecological valency of many species permits transmigration — would corroborate MANN's (1968) description of a basimontane thornbush and dry savanna extending from the Ecuadorian west coast to the east side of the Andes in Peru.

The evaluation of the ranges of the following monotypic species with a common distribution in the area under investigation led to the establishment of the Yungas centre (fig. 21) extending from Cochabamba in Bolivia from 17° south latitude to 6° south latitude in Peru:

1. Manduca reducta
2. Manduca andicola
3. Manduca stuarti
4. Euryglottis guttiventris
5. Sphinx arthuri
6. Sphinx aurigutta
7. Protambulyx ockendeni
8. Amplypterus dentoni
9. Callionima acuta
10. Aleuron cymographum
11. Pachygonia martini
12. Nyceryx maxwelli
13. Perigonia grisea
14. Eupyrrhoglossum corvus
15. Xylophanes docilis
16. Xylophanes ockendeni
17. Xylophanes rhodotus
18. Xylophanes nabuchodonosor
19. Xylophanes pyrrhus

Xylophanes pyrrhus is known to occur sporadically also in Colombia, its principal centre of distribution, however, can be defined in the Yungas centre on the basis of 218 recorded specimens (fig. 52, p. 160). From literature, *Nyceryx maxwelli* is also known to occur in Venezuela. *Manduca albolineata, Manduca chinchilla, Manduca centrosplendens, Neogene steinbachi, Nyceryx draudti, Nyceryx clarki, Aellopos gehleni, Xylophanes caissa* and *Xylophanes sericus* are also said to occur in the East Andean mountane forests of Bolivia and Peru, but they are known to me only from one or few specimens or from literature.

The following subspecifically differentiated sphingids are further elements of of the Yungas centre:

1. Manduca sexta peruviana
2. Manduca diffissa zischkai
3. Manduca scutata boliviana

94

4. Euryglottis albostigmata basalis
5. Isognathus rimosus occidentalis
6. Aleuron neglectum leo
7. Pachygonia ribbei peruviana
8. Nyceryx nictitans saturata
9. Nyceryx continua cratera
10. Enyo bathus otiosus
11. Xylophanes amadis stuarti
12. Xylophanes rothschildi fassli
13. Xylophanes rothschildi bilineata
14. Xylophanes cosmius cosmius

Among them, *manduca sexta peruviana, Xylophanes rothschildi fassli* and *Xylophanes rothschildi bilineata* are not sufficiently documented. Besides, their status is possibly doubtful.

A concentration of small sphingid ranges within the Yungas centre is seen near Cochabamba in Bolivia, on the one hand, and in the Chanchamayo area in Peru, on the other, a fact which cannot alone be explained by especially intensive collecting activities in these areas. Possibly, these areas represent subcentres. The Cochabamba subcentre situated within the provinces of Cochabamba and La Paz has, with a total of 12 sphingids exclusively known from there, a much higher validity as a centre of distribution than the remaining Yungas centre.

From a possible Chanchamayo subcentre, 5 sphingids are known to inhabit this area exclusively.

The evaluation of one single family of Lepidoptera already proves the extreme abundance of taxa in the Yungas centre with a total of 186 different sphingids occurring in this centre, 20 percent of which have not yet spread far beyond the centre.

The observations made on the study of sphingids concur with the statements of other authors. Thus, FORSTER (1958) says about the Yungas fauna: 'It comprises a large number of species specific of this area'. 'This fauna ... might well be one of the most interesting faunas of South America and is certainly the richest in species'.*

11. The Tucuman centre

According to HUECK (1966) who supplies a comprehensive description of the forest regions of the East Andean slopes, the 'laurel forest' in the Tucuman-Bolivian forest area can be regarded as a continuation of the Yungas of Bolivia. It begins in the province of Tucuman normally directly at the foot of the mountains and reaches heights between 600 and 900 m. The sphingids collected in this area occur in these altitudes. The majority of the localities lie between 450 and 750 m

---

* 'Sie umfaßt eine große Anzahl nur ihr eigentümlicher Arten'. 'Diese Fauna ... dürfte wohl eine der interessantesten und wohl die artenreichste Südamerikas sein'.

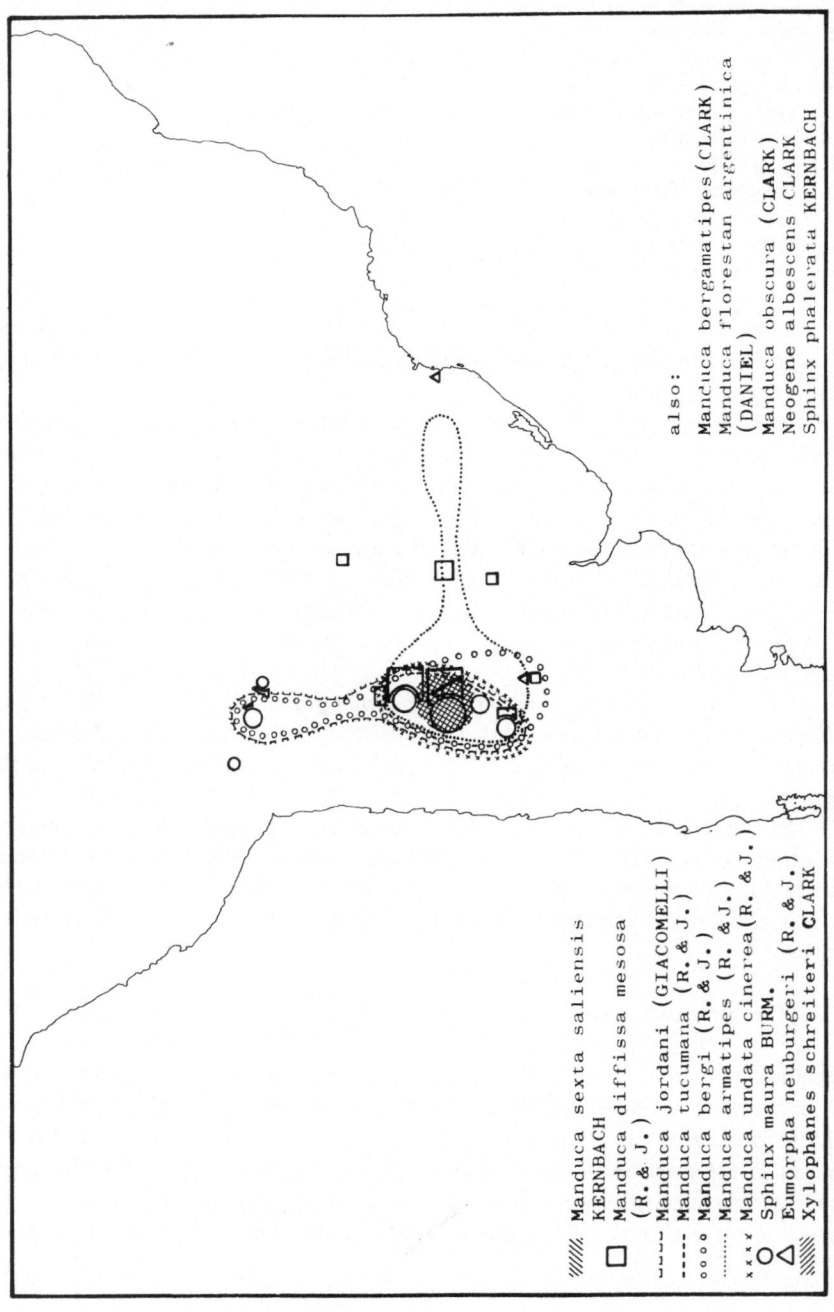

Fig. 22. Faunal elements of the Tucuman centre.

(fig. 51, p. 159). It must, however, be said by way of qualification that many data referred just to the province of Tucuman.

The mapping of the ranges of sphingids known from this area clearly brings forth a nuclear area in Tucuman.

The Tucuman centre represents the southernmost of the Andean centres. The ranges of its elements are limited by decreasing temperatures in the south and increasing altitudes and, in the east, by the dry limit which, according to HUECK (1966), is defined by the course of the 800 mm precipitation curve.

It has already been stressed that the montane rain forest is interrupted south of Santa Cruz, as shown from the vegetation mapping in SCHMITHÜSEN (1969). These factors which prevent spreading in any direction explain the compactness of this centre.

The following monotypic sphingid species with small ranges can be considered as faunal elements of the Tucuman centre:

1. Manduca jordani
2. Manduca tucumana
3. Manduca bergi
4. Manduca bergamatipes
5. Manduca armatipes
6. Manduca carrerasi
7. Manduca obscura
8. Neogene albescens
9. Sphinx maura
10. Sphinx phalerata
11. Orecta acuminata
12. Eumorpha neuburgeri
13. Xylophanes schreiteri

Orecta acuminata was recorded only once, *Sphinx phalerata* from two specimens. The present distribution of the following sphingids indicates that these taxa have undergone their subspecific differentiation in the Tucuman centre:

1. Manduca sexta saliensis
2. Manduca lucetius panaquire
3. Manduca diffissa mesosa
4. Manduca florestan argentinica
5. Manduca undata cinerea

In his catalogue of Argentine sphingids, ORFILA (1933) listed 24 genera with 83 species, 15 subspecies and 2 forms. I found 88 different species and subspecies for the whole of Argentina, a figure that would rise to 108, if strays were included. Of these, 45 (another 34 are insufficiently documented) occur within the Tucuman centre. The area from La Rioja down to the south of Jujuy comprises the actual nuclear area and represents the area of congruence of the evaluated ranges of monotypic species.

Some species show an extension of their ranges as far as the Cochabamba subcentre (fig. 22), others have spread via Paraguay in the direction of the Serra do Mar. *Neogene reevi* has an approximately equal distribution in the Tucuman centre and in the Serra do Mar. *Callionima grisescens* (fig. 56, p. 164) seems to have spread northward from the Tucuman centre, its original main centre of distribution. As for the specimens known from Venezuela, it has to be examined whether they belong to the subspecies *Callionima grisescens elegans*, and whether this subspecies described by GEHLEN (1935) from Pernambuco is valid.

It is conspicuous that there are few species with a wide range of distribution in the Andes which extends beyond the Yungas centre to Tucuman. Nevertheless, disjunct patterns of occurrence in the Andes, on the one hand, and in the Serra do Mar, on the other, are frequent. Even well documented species with wide ranges of distribution, such as *Protambulyx strigilis, Amplypterus gannascus, Pseudosphinx tetrio* and *Callionima parce* (figs. 57, 58, 59, 60, pp. 165-168) have their southernmost distribution limit on the Andean side north of the Tucuman centre, leaving strays apart. From all this it appears that, in the Tucuman centre, specific conditions prevail which prevent many sphingids from living in this area. The species, therefore, which are still able to tolerate these conditions, occupy an exceptional position which is expressed by a common distribution centre.

The distribution pattern of sphingids in Ecuador and Colombia

In contrast to the Tucuman centre which is characterized by its compactness, a comparative chorological investigation of the sphingids found in the North Andes brings forth a complex distribution pattern. One of the reasons for this is that different orographic conditions prevail in this area. Besides the coastal chain (300 to 600 m), there are two 4,000 to 5,000 metres high mountain chains in Ecuador, and in Colombia, there are 3 (including the coastal cordillera 4) meridionally running mountain ranges with large longitudinal valleys extending in between.

As a result of the drifting of the cold HUMBOLDT Current off the coast in the Guayaquil region, both the northern part of the coastal plane of Ecuador and the Pacific coast of Colombia are characterized by a humid tropical climate. HUECK (1966) said that this is one of the geographical zones with the greatest rainfall. This also explains the occurrence of tropical lowland rain forests on the Pacific coast. The rain forests extend northward to Central America. In the south of Ecuador, on the other hand, the zone of transition 'from the humid-tropical rain forest to the thornbush savanna of the semi-desert' is said to cover a distance of as little as 200 km (Meyers Kontinente und Meere 1969). The west slopes of the Andes from approximately 4° south latitude are covered with montane rain forest, as are the East Andean slopes (cf. SCHMITHÜSEN 1969). The valleys of the Rio Cauca and the Rio Magdalena are said to be relatively dry, due to their position on the leeward side of the Cordilleras.

In accordance with the occurrence of montane rain forest on either side of the Andes, frequent records of sphingids exist also from the Pacific side and the Cauca valley, starting from Central Ecuador northward (fig. 51, p. 159). Localities from the intermontane depressions of Ecuador have been registered either on the left or on the right according to the opening these depressions show either to the east or the west by the rivers draining them. In Colombia, a separation between eastern and western localities on either side of the central cordillera has been made.

In Ecuador just as in Colombia, there occur altogether 159 specifically and subspecifically differentiated sphingids, of which 15 from Ecuador and 24 from Colombia are insufficiently documented. While, in Peru, the Andes proved impenetrable to the montane forest species, apart from some exceptions (MOSS 1912), and, in addition, the necessary ecological conditions are lacking on the Pacific side, in the north the Andes are more and more losing their importance for sphingids as a recently effective isolation barrier. A comparison between the number of sphingids occurring in the eastern and western parts of Ecuador and Colombia can substantiate this observation: of a total of 200 different sphingids confirmed from Ecuador and Colombia, 112 species and subspecies were recorded from East Ecuador; 52 also from the west, 19 of which, however, are not represented in the east of Ecuador.

In Colombia, 112 species and subspecies occur in the east, and almost just as many, altogether 111, also in the west, including 21 which have not so far been proved to occur in the eastern part of Colombia. The greater permeability of the northern Andes as well as the possibility for sphingids to live also on the Pacific side and in the intermontane depressions and river valleys result in a complex distribution pattern which does not permit the establishment of a distinct centre of distribution. However, it can be assumed that, in this area, several centres must have existed which served both as areas of refuge and centres of dispersal.

HAFFER (1967) referring in particular to the West Colombian lowland rain forest said that it is 'characterized by a surprisingly large number of endemic birds'. He found a 'Choco-Refuge' confirmed on the Pacific coast of Colombia. MÜLLER (1970, 1973) studying the ranges of vertebrates found a Colombian montane forest centre, a Colombian Pacific centre, a North Andean centre as well as centres in the Cauca and Magdalena valleys.

The mapping of the ranges of sphingids with a wide distribution in the Andes shows a blank in the south of Colombia (figs. 45, 46, 49, pp. 153, 154, 157). The south of Ecuador is still reached by species coming from the Yungas centre. Sphingids occurring only in Ecuador are *Cocytius macasensis, Manduca camposi, Sphinx merops judsoni, Nyceryx lunaris, Eumorpha macasensis, Xylophanes agilis* and *Xylophanes macasensis*, which, however, with the exception of *Nyceryx lunaris*, have so far been insufficiently documented. *Nyceryx lunaris* occurs on the East Andean side. *Euryglottis davidianus* and *Xylophanes dolius* also occur in the East Andes; but besides in Ecuador, they are also found in northern Peru. The monotypic genus *Protaleuron rhodogaster* has been recorded by one specimen each in Ecuador and Peru. *Eumorpha drucei* occurs in the west of Ecuador and in the Cauca valley in Colombia. The range of *Euryglottis albostigmata* is made up by

* 'vom feucht-tropischen Regenwald zur Dornstrauchsavanne der Halbwüste'

localities in East Ecuador and West Colombia from the Rio Dagua and the Cauca valley. With the exception of *Xylophanes mirabilis*, the following species and subspecies have been found confirmed for the Pacific side of Colombia only by inaccurate locality data such as 'West Cordillera' or 'West Colombia':

1. Sphinx merops monjena
2. Protambulyx goeldii andicus
3. Aellopos titan aguacana
4. Xylophanes hojeda
5. Xylophanes fusimacula niepelti
6. Xylophanes mirabilis
7. Xylophanes colombiana

The ranges of sphingids as they present themselves in Ecuador and Colombia do not so far justify the establishment of a definite centre of distribution in these areas. On the one hand, this might be explained by the possibilities the sphingids had to disperse in the northern Andes; on the other hand, the species with small ranges described from these areas, which might qualify as indicators for possible centres in Ecuador, in the region of the West Cordilleras of Colombia or in the Cauca valley, have so far been confirmed only insufficiently.

12. The Venezuelan centre

Before entering Venezuela, the Colombian East Cordillera of the Andes forks into a mountain range which extends as the Sierra de Perija west of Lake Maracaibo, where it forms the border between Colombia and Venezuela, and into the Sierra Nevada de Mérida, which continues into the Caribbean Coastal Cordillera and east of the depressions of the Rio Unare and Rio Aragua de Barcelona into the mountainous country of Sucre.

According to HUECK (1966) the easy access to the montane forests of the Cordillera de Mérida, of the Caribbean Coastal Cordillera and the Macizo Oriental has permitted man's influence to become very pronounced everywhere, except in the Sierra de Perija which is said to be completely untouched to this day.

This statement concurs with the results of sphingid collecting in these areas. Even of widespread species, locality data are lacking from the Sierra de Perija. The ranges of species with a wide pattern of distribution in the Andes, like *Manduca scutata* (fig. 45, p. 153) extend, in Venezuela, beyond the Cordillera de Mérida up to the Caribbean coast. In the whole of Venezuela, 166 sphingid species and subspecies were counted, 16 taxa of which are poorly known.

The sphingids of Venezuela have been especially well studied by LICHY (1943-1968). Furthermore, the following authors have reported on sphingids from Venezuela: BEEBE & FLEMING (1945), CARY (1949), DANIEL (1949b), FLEMING (1947), GEHLEN (1935), KERNBACH (1965b), VOGL (1944). The subsequently listed sphingids are confirmed from Venezuela only. They indicate a centre of distribution which is located in the Caribbean Coastal Cordillera:

100

1. Manduca empusa
2. Manduca maricina
3. Manduca florestan vogli
4. Isognathus rimosus papayae
5. Perigonia pittieri
6. Enyo taedium australis
7. Xylophanes germen yurakano
8. Xylophanes amadis meridianus

Of the monotypic species, *Manduca maricina* was recorded only from literature. Another monotypic species from Venezuela is *Isognathus tepuyensis*. It has been described by LICHY (1962) as inhabiting the Guyana Highlands of Venezuela.

Callionima calliomenae (fig. 61, p. 169) has its main distribution in Venezuela, from where it extens to Colombia and also to Hispaniola. *Manduca dilucida, Kloneus babayaga, Xylophanes neptolemus* and *Xylophanes tyndarus* are species which have a share in Central America and Venezuela.

A subspecies *trinitatis* of *Xylophanes neptolemus* has been described from Trinidad; if valid, it would be the only endemic sphingid occurring on this island.

On the basis of vertebrate ranges MÜLLER (1970) could establish a coastal forest centre and a montane forest centre within the Caribbean Coastal Cordillera. The ranges of sphingids do not permit a similar differentiation; the evaluation of the mentioned species with small ranges, however, led to a common nuclear area whose position in the Caribbean Coastal Cordillera coincides with the two centres established by MÜLLER(1970).

Among the subspecies pertaining to this centre, *Isognathus rimosus papayae* and *Xylophanes amadis meridianus* are also strongly represented in the Cordillera de Mérida, and it cannot be excluded with certainty that this has been the site where their subspecific differentiation took place.

In the southern part of the Cordillera de Mérida, a gap of localities of occurrence is conspicuous, which, according to the vegetation mapping in SCHMITHÜSEN (1969), could be ecologically explained by an interruption of the tropical montane rain forests in this area.

The remaining South American centres

13. The Serra do Mar centre

The rain forest biomes of the Brazilian coastal range are obviously characterized by conditions very similar to those of the montane rain forests of the Andes. A large number of sphingids, such as *Perigonia stulta* and *Xylophanes titana* (fig. 62, p. 170), show a disjunct distribution in the Andes or beyond the Andes in Central America, on the one hand, and in the Serra do Mar, on the other. Other species show this affinity through closely related subspecies, such as *Manduca pellenia pellenia* and *Manduca pellenia janeira* or *Nyceryx nictitans nictitans* and *Nyceryx nictitans saturata* (fig. 63, p. 171). The distribution pattern of *Phryxus caicus* within

101

South America, however, rather shows a disjunct distribution between the lowland rain forests of Guiana and Amazonia and the Serra do Mar (fig. 30, p. 138).

Within the Brazilian coastal range, the sphingid ranges permit to establish a centre of distribution extending from approximately 22° to 28° south latitude. This centre includes the Serra do Mar and the Serra da Mantiqueira which runs parallel in the north, as well as a small part of the Serra Geral in the south.

MAACK (1969), discussing the different views of the definition of the term 'Serra do Mar' as against the great number of locally used Serra-names, defined the Serra do Mar as the 'large eastern border chain of the Brazilian High Plateau ... extending from the State of Rio de Janeiro to Santa Catarina'.*

According to MAACK (1969), the Serra do Mar is covered with dense rain forests from the coastal plain up to 1,100 m. The abundant vegetation results from the high precipitation brought along all over the Brazilian east coast the whole year round by the southeast trade wind (Meyers Kontinente und Meere 1969, HUECK 1966).

The majority of localities of sphingids with small ranges within the Brazilian coastal range are concentrated on the Atlantic side. There are also species, however, which penetrate more or less far into the west, such as *Amplypterus eurysthenes* or *Manduca incisa* (fig. 64, 65, p. 172, 173).

The decisive isolation factors for sphingids might not so much be seen in the 1,500 to 2,300 m high altitudes of the coastal range, but rather in the more arid climatic conditions prevailing in the west. 'In the surroundings of Curitiba and west of the Serra do Mar Campos and Araucaria forests extend' (MAACK 1969).**

In the Serra do Mar, 177 specifically and subspecifically differentiated sphingids were recorded, 36 of which are insufficiently documented. As compared to this, HOFFMANN (1934) listed 119 species for Santa Catarina. As faunal elements of the Serra do Mar centre, we first have to consider the following monotypic species which are either restricted, with their small ranges, to the centre in question or have their definite main centre of distribution in this area:

1.	Manduca grandis	12.	Aleuron prominens
2.	Manduca suavis	13.	Nyceryx nephus
3.	Sphinx justiciae	14.	Eumorpha translineata
4.	Protambulyx fasciatus	15.	Xylophanes depuiseti
5.	Protambulyx astygonus	16.	Xylophanes xylobotes
6.	Amplypterus eurysthenes	17.	Xylophanes indistincta
7.	Amplypterus germanus	18.	Xylophanes fosteri
8.	Isognathus australis	19.	Xylophanes ferotinus
9.	Hemeroplanes longistriga	20.	Xylophanes kaempferi
10.	Callionima parthenope	21.	Xylophanes hydrata
11.	Aleuron ypanemae	22.	Xylophanes aglaor

* 'die große östliche Randstufe des brasilianischen Hochplateaus ... vom Staat Rio de Janeiro bis nach Santa Catarina'.
** 'In der Umgebung von Curitiba und westlich der Serra do Mar erstrecken sich Campos (Grasfluren) und Araukarienwälder'.

Of the mentioned species, *Manduca grandis, Protambulyx fasciatus, Amplypterus germanus* and *Xylophanes ferotinus* were recorded only on the basis of one specimen each or taken from literature.

The existence of the Serra do Mar centre can be further substantiated by the following sphingids subspecifically differentiated in this centre:

1. Manduca diffissa petuniae
2. Manduca hannibal hamilcar
3. Manduca pellenia janeira
4. Manduca scutata brasiliensis
5. Manduca incisa incisa
6. Manduca dalica anthina
7. Protambulyx xanthus australis
8. Orecta lycidas lycidas
9. Isognathus rimosus brasiliensis
10. Callionima pan neivae
11. Nyceryx nictitans nictitans
12. Nyceryx continua continua
13. Nyceryx alophus alophus
14. Enyo japix discrepans
15. Enyo pronoe fuscatus
16. Aellopos fadus flavosignata
17. Eumorpha obliquus orientalis
18. Xylophanes tyndarus marginalis
19. Xylophanes schausi schausi
20. Xylophanes isaon isaon

Aellopos fadus flavosignata was confirmed by one specimen only. *Manduca hannibal hamilcar* is possibly doubtful.

In comparison with the remaining Neotropical centres, the Serra do Mar centre makes itself conspicuous by the highest number of sphingids that can be recognized as faunal elements of any of the centres. The reason for this might be both the favourable ecological conditions prevailing for sphingids in the Serra do Mar and the quality of the surrounding isolation barriers.

It seems especially the Campo Cerrado which has a particular impeding effect on the spreading of sphingids mainly adapted to rain forest biomes, as seen from the practically complete lack of localities even in the case of widespread species.

On the basis of vertebrate ranges MÜLLER (1970) defined the position of a Serra do Mar centre extending on the east side of the Brazilian coastal range from the State of Santa Catarina in the south to the state of Pernambuco in the north, including the continental islands offshore these states. He could establish three subcentres, of which the southernmost 'Paulista subcentre' – which MÜLLER calls the most important of the three subcentres – coincides with the Serra do Mar centre defined on sphingid ranges (fig. 23).

None of the small-range indicator species extends beyond 20° south latitude in the north.

Most of the species have their northern limits in the region of Cabo Frio. This observation is in conformity with HUECK's (1966) description of an interruption of the coastal rain forest by a long arid zone (cf. MÜLLER 1970, 1973). Only very widespread species, such as *Protambulyx strigilis* or *Pseudosphinx tetrio* (figs. 57, 59, pp. 165, 167) inhabit both the Serra do Mar and the north of the Brazilian coastal range. There are no small-range species, however, which would corroborate such a connection.

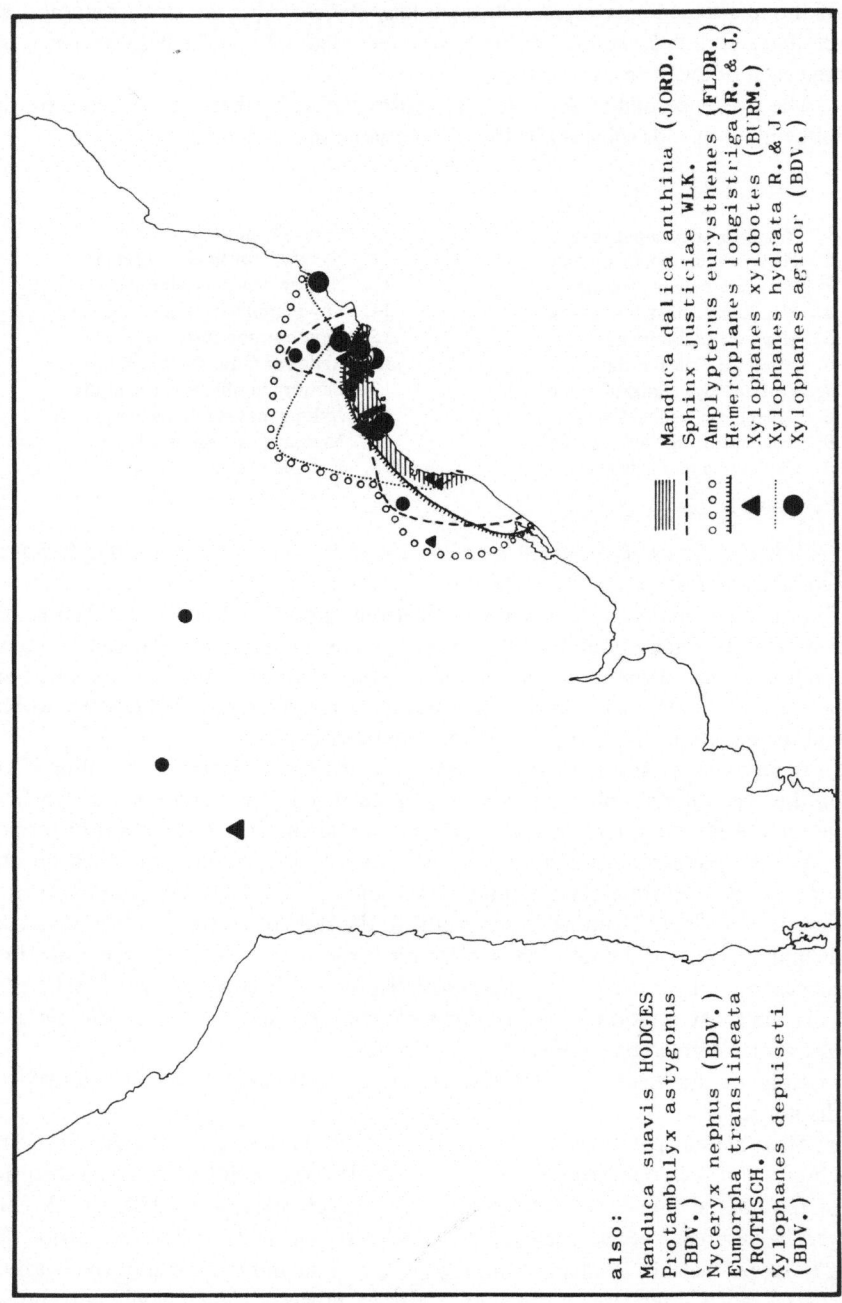

Fig. 23. Faunal elements of the Serra do Mar centre.

Isognathus allamandae, as a monotypic species, *Neogene dynaeus* and *Callionima grisescens elegans* in its subspecific differentiation are almost exclusively restricted to Pernambuco. These elements could be considered as indicators of a possible Pernambuco centre.

14. The Uruguayan centre

The Uruguayan centre defined on the ranges of vertebrates can be substantiated by two monotypic sphingid species, i.e. *Manduca feronia* and *Xylophanes alegrensis* which have been described from the border area between Uruguay and South Brazil. In addition, the following subspecifically differentiated sphingids show a distribution pattern which allows them to be considered as faunal elements of a possible Uruguayan centre:

1. Manduca lucetius exiguus
2. Manduca diffissa diffissa
3. Orecta lycidas eos
4. Nyceryx alophus ixion
5. Hyles lineata lineatoides

One other subspecies each of *Orecta lycidas* and *Nyceryx alophus* occur in the adjacent Serra do Mar centre, which would speak in favour of an Uruguayan centre (fig. 24). The subspecies *mesosa* of *Manduca diffissa* has been found to be a faunal element of the Tucuman centre.

BIEZANKO & ZOPP (1954), however, question the occurrence of *Orecta lycidas eos* in Uruguay. The localities recorded in six different museums read: Montevideo, Rio Uruguay, Buenos Aires and La Plata. Other less precisely defined localities are Paraguay and Argentina. The majority of localities are known from Buenos Aires.

Manduca lucetius and *Hyles lineata* are claimed by BIEZANKO & ZOPP to occur in Uruguay in the nominate forms, whereas *Manduca diffissa* is said to inhabit Uruguay with its subspecies *petuniae*. The low number of indicator species is explained from the decreasingly favourable living conditions for sphingids in this area. In Uruguay, a total of 34 specifically and subspecifically differentiated sphingids occur. Another 10 sphingids were recorded in one specimen each.

Starting from Rio Grande do Sul which partly belongs to the Uruguayan centre (MÜLLER 1970, 1973), a change takes place from the rain forest to the open grassland (HUECK 1966). According to RAMBO (1958), the rain forest is now in an expansion phase southwards. In Uruguay, subtropical humid forest occurs at least along the rivers, and in the valley of the river Uruguay this forest reaches widths up to 10 km (Meyers Kontinente und Meere 1969). MÜLLER (1970) said that 'faunal elements of the Serra do Mar centre penetrate into the Uruguayan centre along gallery forests'.*

* 'Entlang von Galeriewäldern dringen Faunenelemente des Serra-do-Mar-Zentrums in das Uruguay-Zentrum vor'.

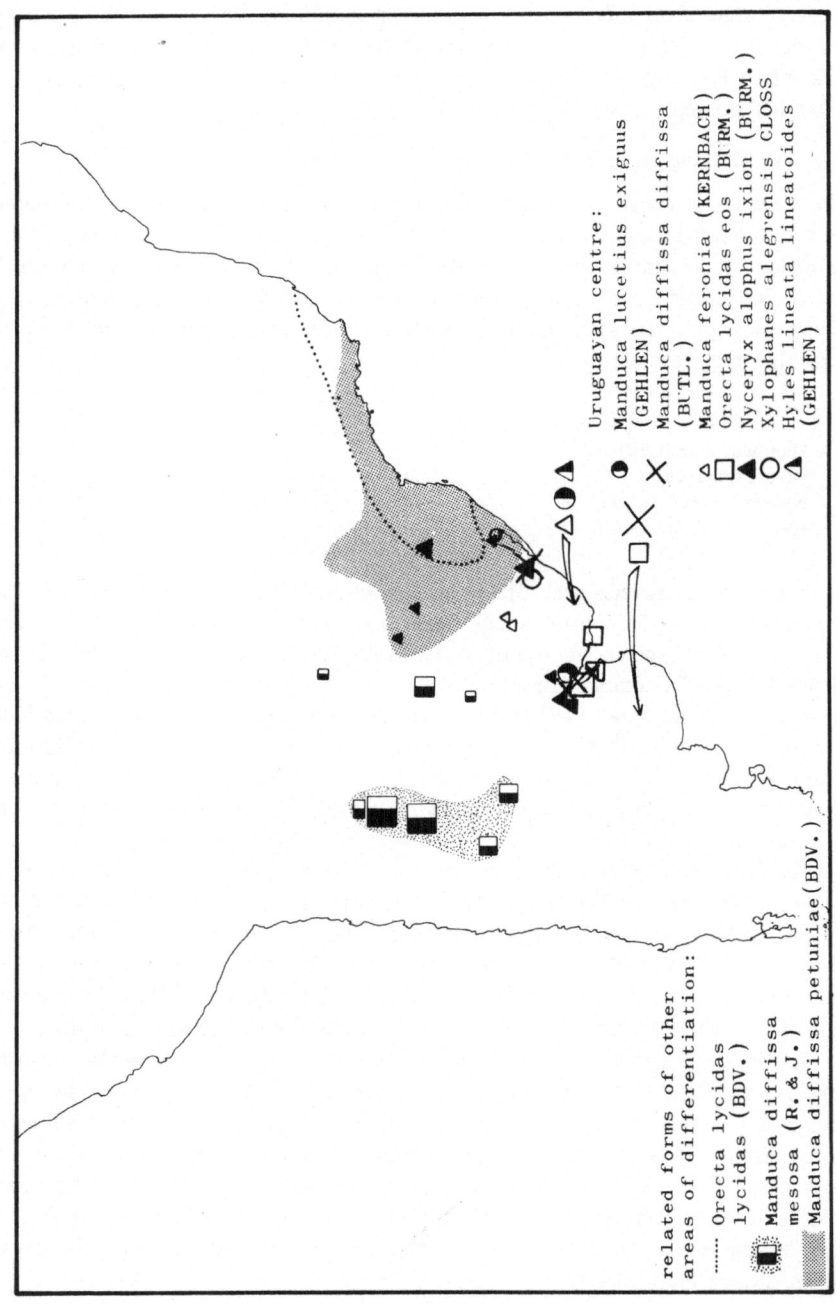

Fig. 24. Relations between southern Neotropical dispersal centres.

Under more arid climatic conditions, this might have led to the isolation and subspecific differentiation of species such as *Orecta lycidas* and *Nyceryx alophus*.

15. The Andean Pacific centre

The ecological conditions prevailing on the Pacific coast of Peru and Chile are unfavourable for sphingids especially due to the influence of the HUMBOLDT Current and the coastal desert it has created.

URETA & DONOSO (1956) say about Chile: 'Nuestro pais se ha caracterizado por la escasez de species de esta notable familia ...' In their revision of the sphingids of Chile, they mention the following 15 species:

 1. Agrius cingulatus
 2. Cocytius antaeus medor as Cocytius medor
 3. Manduca sexta caestri
 4. Manduca stuarti
 5. Manduca rustica
 6. Sphinx aurigutta
 7. Pseudosphinx tetrio
 8. Erinnyis ello
 9. Aellopos tantalus as Sesia tantalus eumelas
10. Eumorpha satellitia
11. Eumorpha labruscae
12. Xylophanes tersa
13. Hyles euphorbiarum
14. Hyles annei
15. Hyles lineata

In addition to these, *Manduca lucetius panaquire* and *Pachylia ficus* were recorded for Chile.

The conditions prevailing on the Pacific side of Peru are similar. MOSS (1912) found 19 species on the west coast during his three years' stay in Lima; as compared to this, JORDAN listed approximately 120 species for the total of Peru in his introduction to the same work. I found 151 sphingid species and subspecies confirmed in Peru and another 36 with poor records.

MOSS says that, of the species he found *Pseudosphinx tetrio* and *Eumorpha labruscae* do not represent coastal species, but immigrants (i) from the interior of the country which must definitely have come from the other side of the Andes, where the plants occur on which they feed. He assumes that *Erinnyis lassauxi*, *Erinnyis crameri* and *Erinnyis obscura* are likewise immigrants from the interior of the country. In Lima and the remaining coastal area in which he collected, MOSS found the following species:

1. Agrius cingulatus
2. Cocytius antaeus medor
3. Coytius lucifer
4. Manduca sexta paphus

	5.	Manduca mossi
	6.	Manduca rustica
	7.	Euryglottis davidianus
(i)	8.	Pseudosphinx tetrio
(i)	9.	Erinnyis lassauxi
	10.	Erinnyis ello
(i)	11.	Erinnyis crameri
(i)	12.	Erinnyis obscura
	13.	Pachylia ficus
	14.	Eumorpha vitis
	15.	Eumorpha fasciata
(i)	16.	Eumorpha labruscae
	17.	Xylophanes tersa
	18.	Hyles annei
	19.	Hyles lineata

Manduca mossi is a monotypic endemic species of Lima and its surroundings and occurs in altitudes up to 7,000 feet.

Euryglottis davidianus was considered by MOSS to be restricted to the West Andean slopes between 7,000 and 8,000 feet. This species has meanwhile also been found in the southeastern part of Ecuador.

GEHLEN (1942) described *Manduca chinchilla* from the southwest of Peru.

Manduca sexta caestri shows a distribution pattern almost exclusively restricted to Chile (fig. 25).

Hyles annei, on the other hand, inhabits the coast from Valpareiso up to Ecuador.

The Andean Pacific centre can be defined on the area of congruence of the ranges of the aforementioned species. It is situated within the Andean Pacific centre established by MÜLLER (1970) on the basis of vertebrate studies.

16. The Paraguayan centre

The mapping of sphingid ranges in Paraguay indicates a possible centre of distribution between the Rio Paraguay and the Rio Parana.

The vegetation mapping in SCHMITHÜSEN (1969) indicates mainly subtropical rain forest for this area.

In Paraguay, a total of 86 different sphingids were recorded, 12 of which were poorly documented.

In East Paraguay, the following sphingids occur:

1. Neogene reevi minor
2. Neogene pictus
3. Neogene intermedia
4. Aleuron neglectum paraguayana
5. Enyo gorgon heinrichi
6. Xylophanes anubus paraguayensis

108

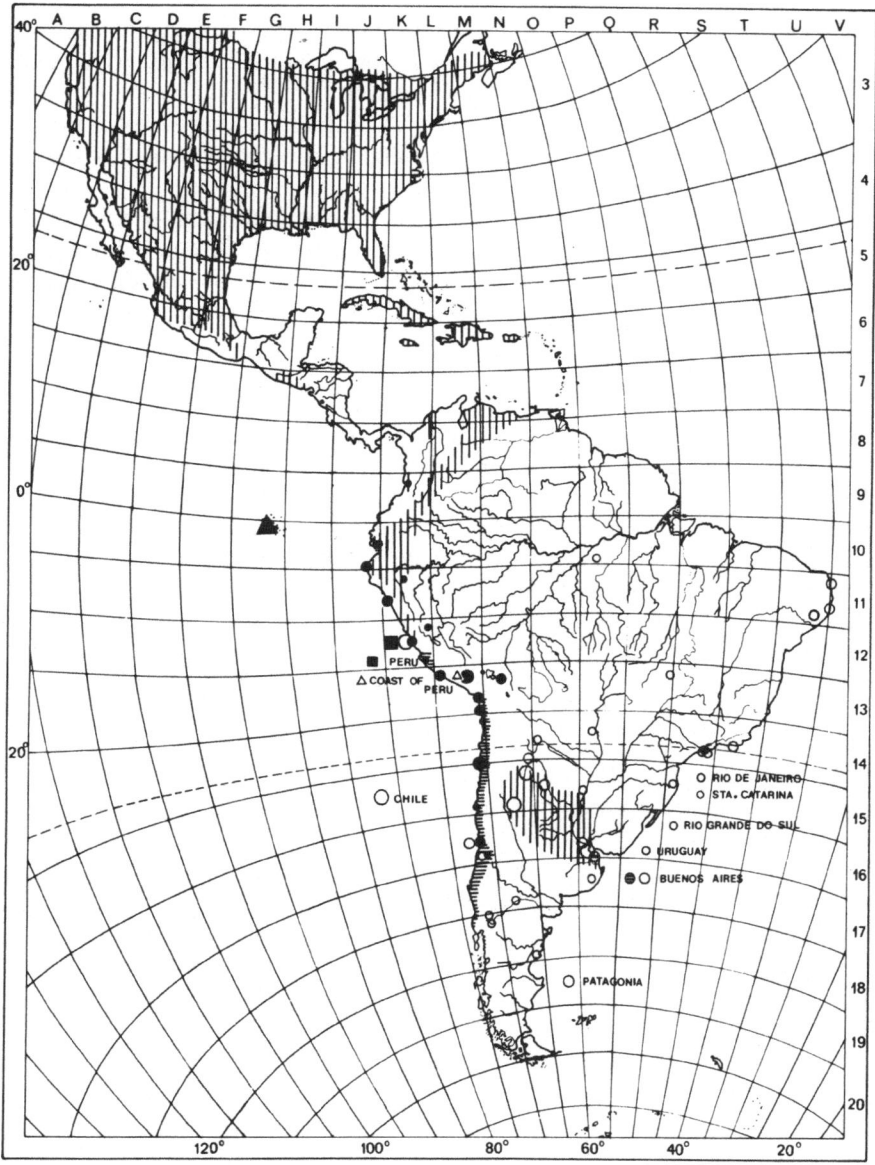

Fig. 25. Faunal elements of the Andean Pacific centre and distribution of related taxa of *Hyles annei* (GUÉR.) ●: *Manduca sexta caestri* (BLCH.) ≡, *Manduca mossi* (JORD.) ■, *Manduca chinchilla* (GEHLEN) △, *Hyles lineata lineata* (F.) ‖‖, *Hyles lineata florilega* (KERNBACH) ▲, *Hyles euphorbiarum* (GUÉR. & PERCH.) ○.

Neogene pictus and *Neogene intermedia* are monotypic species.

A relationship between the Paraguayan centre and the Mato Grosso centre is suggested by *Manduca manducoides* and *Periogonia lusca passerina* which have their main distribution in the Mato Grosso centre.

Nyceryx alophus ixion from the Uruguayan centre and *Protambulyx astygonus* as well as *Xylophanes fosteri* from the Serra do Mar penetrate into the Paraguayan centre.

In East Paraguay, a centre of distribution has not so far been established. In view of the fact that, if one family only is investigated, the yield of indicators for a centre is, sometimes, quite restricted, I would rather deduce only the probable existence of a Paraguayan centre from the present studies of sphingid ranges.

17. The Mato Grosso centre

The ranges of sphingids indicate another centre of distribution in the south-western Mato Grosso area on the upper river Paraguay. The nuclear area is located between Cuiabá, Mato Grosso and Corumbá in the humid savannas of the Pantanal flooded during the rainy seasons (HUECK 1966). SCHMITHÜSEN (1969) mapped for this area tropical rain-green monsoon forests, as they also occur, scattered like islands in the Campo Cerrado, on the river Paraná and between the Rio Paranaiba and the Rio Grande. The localities of *Manduca corumbensis* and *Neogene curitiba* (fig. 66, p. 174) correspond to these forest islands.

By evaluating the ranges of vertebrates, MÜLLER (1970) could establish a Campo Cerrado centre. He pointed out, however, that the Campo Cerrado has never been completely unforested and that it obviously represents a zone of interpenetration of grassland and forests.

As for Sphingidae it can be assumed that they rather inhabit the scattered forest biotopes.

The fact that the number of species (totalling 70 species and subspecies) occurring in this area as compared to those in the Andes and the Serra do Mar is much lower, reflects the much more unfavourable ecological conditions prevailing here for sphingids. On the other hand, it has to be pointed out that in this relatively thinly populated area (KÜHLHORN 1959), collecting activities have not yet been carried out to any comparable extent.

The following species and subspecies are to be reckoned as faunal elements of the Mato Grosso centre:

1. Manduca manducoides
2. Manduca viola-alba
3. Manduca brunalba
4. Manduca albiplaga exacta
5. Manduca corumbensis
6. Neogene curitiba
7. Neogene dynaeus corumbensis
8. Callionima modesta

9. Perigonia pallida rufescens
10. Perigonia lusca passerina
11. Eumorpha satellitia excessus

Of these, 6 species are monotypic. The status of *Manduca albiplaga exacta* is doubtful, since the nominate form occurs in Central America and in the Andes on the one hand and in the Serra do Mar on the other.

Possible further faunal elements of the Mato Grosso centre are *Manduca incisa pallidula* and *Perigonia leucopus*; the former is, however, also known from Bolivia and Mexico, the latter from the Serra do Mar, the Andes and, in one specimen, also from Hispaniola.

18. The Guiana centre

On the basis of vertebrate ranges, MÜLLER (1970) could establish a 'Guyanan' centre extending from the Orinoco delta southward almost as far as the Amazon. As for Pará, MÜLLER found an individual rain-forest centre confirmed.

The ranges of some sphingid species which are to be regarded as faunal elements of the Guiana centre extend beyond the Amazon as far as Pará. Many of the locality data from Pará refer to Belém (formerly Pará), where MOSS (cf. MOSS 1920) collected sphingids in the city of Belém and its surroundings. He found a total of 90 species for Pará. I recorded 96 different sphingids from Pará, 9 of which have been confirmed by one specimen each only. From Guiana 115 sphingids were recorded, 15 only once. BEEBE & FLEMING (1945) found 53 species in Kartabo, Guyana, in an area no larger than quarter of a square mile, where they have been collecting at intervals during a total of eight years.

These authors pointed to the basic similarity between Pará and Kartabo with regard to their position in the heart of the tropical climax rain forest with its typical lack of alternating humid and dry seasons.

The relative homogeneity of environmental conditions in the Guiana states and Pará might be the reason for the common occurrence of sphingids in these parts of South America.

The following sphingids are faunal elements of the Guiana centre:

1. Cocytius lucifer lindneri
2. Manduca vestalis
3. Isognathus scyron
4. Xylophanes ploetzi
5. Xylophanes mossi
6. Xylophanes epaphus
7. Xylophanes guianensis
8. Xylophanes amadis amadis
9. Xylophanes amadis goeldi

Six species are monotypic.

Cocytius lucifer lindneri has been described by GEHLEN (1944) on the basis of one male specimen only. The range of *Isognathus sacryon* (fig. 67, p. 175) extends from Central America to Venezuela and the Guiana states as far as Pará and Manaus. *Xylophanes ploetzi* is lacking in Pará. *Xylophanes amadis* occurs in its nominate form in Guiana, whereas the subspecies *goeldi* inhabits Pará, which would fabour the subdivision of the Guiana centre into two subcentres. Other subspecies of *Xylophanes amadis* have been described from Central America, Venezuela and the Andes.

Besides their occurrence in the Guiana centre, *Xylophanes guianensis* and *Xylophanes epaphus* show a scattered occurrence in the Serra do Mar and the Andes and in Panama respectively.

Xylophanes rufescens appears to have a disjunct distribution in Guiana and Peru; two localities from the Amazon region, namely Fonte Boa and São Paulo de Olivença, might, however, also point to a possible range relationship between these two areas, an assumption which would have to be confirmed by further records from the Hylaea.

The distribution of sphingids in the Hylaea

HUMBOLDT (1859) described, under the term 'Hylaea', the ever-humid tropical rain-forest of the Amazon lowland which, according to NEEF (1968), covers an area of 4.5 million square kilometres and, according to SCHMITHÜSEN (1968), represents the largest uninterrupted area of tropical rain-forest.

The only distinct topographical limits are the Andes in the west and the southern rim of the Guiana Highlands in the north. In the northwest and south, the uninterrupted evergreen rain-forest is bordered by the open grassland of the Campos and Llanos respectively, or by more open deciduous plant associations (NEEF 1968).

The Hylaea is interspersed with a number of forest-free Campo islands whose existence can most probably be brought into connection with climatic changes (HUECK 1966, MÜLLER & SCHMITHÜSEN 1970).

The climate of Amazonia is characterized by stable temperatures throughout the year rather than by extremely high temperatures (HUECK 1966). SIOLI (1968) referring to precipitation, also writes that 'one cannot speak of a real dry season within the Hylaea area'.*

The homogeneity of the ecological conditions is in contrast to the frequently emphasized richness in species (cf. FITTKAU 1969). Lacking isolation barriers within Amazonia are one of the reasons why there are hardly any small sphingid ranges that would make it possible to define the location of a distribution centre. With 76 well established species and 24 others recorded from the Hylaea in one or two specimens only, the total number of sphingid taxa lies far beneath those known from the Andes or Venezuela.

Furthermore, the vast majority of the mentioned species are either widespread or appear to be penetrating from West Amazonia or from the Andes.

* '... im Gebiet der "Hyläa" von einer wirklichen Trockenzeit nicht die Rede sein kann'.

Isognathus mossi is the only monotypic species to be known from the Amazon exclusively, mainly from the Manaus area, with the exception of one specimen which was described from Rio de Janeiro.

Manduca incisa prestoni, Enyo cavifer paganus and *Xylophanes cosmius obscurus* also have their distribution ranges in Central Amazonia, evidence, however, is too scanty to justify the establishment of a definite centre of distribution. *Nyceryx stuarti* and *Xylophanes wolfi* are mainly distributed in the west of the Hylaea, whereas *Cocytius mortuorum, Manduca clarki, Nyceryx magna* and *Phanoxyla hystrix* seem to have spread from the Andes into the Hylaea.

As can be seen from the mapping of localities of widespread species such as *Protambulyx strigilis* and *Pseudosphinx tetrio* (figs. 57, 59, pp. 165, 167) or from the locality guide-map of South America (fig. 69) the evidence of sphingids occurring in the Hylaea is still very incomplete. The majority of the localities of occurrence are found along the large rivers which, until recently, have been the only traffic routes within the tropical rain-forest of the Amazon lowland.

On the other hand, however, the question imposes itself whether the relatively low number of sphingid species occurring in the Hylaea as compared with the Andes or the Serra do Mar can be interpreted as an indication to the dependency of sphingids on montane biomes, which could be explained either historically or by reasons of recent ecology.

VI. DISCUSSION

a. Distribution centres as refuge areas and dispersal centres

The 18 Neotropical distribution centres (fig. 8, p. 66) which could be analysed on the basis of the comparative chorological evaluation of sphingid ranges, are of different valencies with regard to the number of monocentric species or subspecies which could be recognized with certainty as faunal elements of the different centres. With a total of 42 and 33 sphingids respectively with a small-range distribution the Serra do Mar centre and the Yungas centre proved to be the best verifiable centres; they are followed by the Central American rain forest centre with 23 and the Tucuman centre with 18 species and subspecies with small ranges. The centres with the lowest number of small-range sphingids are the Andean Pacific centre in South America and the Yucatán centre in Central America. The possible establishment of additional centres in Ecuador and Colombia or within the Hylaea on the basis of sphingid ranges requires more exhaustive collecting in the areas in question.

By evaluating 4817 South and Central American vertebrate ranges, excluding the Antilles, MÜLLER (1970, 1973) could prove the existence of 40 dispersal centres in the Neotropical region (fig. 9, p. 67). The studied vertebrates included groups with the most varying demands on their habitats, as Amphibians and Reptiles of which, in contrast to the former, many are adapted to more arid conditions.

The proof of both arboreal centres and non-forest centres could be expected from this specific study. Quite obviously, the study of only one family with a total of 518 species and subspecies partly or wholly distributed in the Neotropical region could not lead to the confirmation of all the centres established on the basis of vertebrate ranges. This has different reasons, but in the first place it can be explained by the ecological valency and the vagility of the sphingids.

The great flying capacity by which the spingids are characterized deserves particular attention in a zoogeographical investigation. It cannot, however, be a reason to reject the sphingids as suitable objects for distribution analyses altogether, just as little as this would be justified with birds which have frequently been the objects of similar investigations (HAFFER 1967, 1969, 1970, MAYR & PHELPS 1967, MONROE 1968, MÜLLER 1970, 1973, VUILLEUMIER 1970). It rather speaks in favour of the general validity of the applied method that even within an animal group with great power to disperse there are sufficient static species to allow the identification of the dispersal centres of expansive taxa. The

114

processes of dispersal can be inferred from the falling gradient of the range borders starting from the nuclear areas in the sense of REINIG (1950).

The fact that the ranges of the species with a limited expansion are not spread irregularly over a given region, but can be arranged in centres whose position in many cases shows a congruence with those of completely different organisms investigated with a view to such centres, can only be explained by the effects of outward factors and conditions to which all living creatures of certain areas had been exposed at the same times. Most obviously, such factors must have been macroclimatic changes to which the biota reacted with range regressions or expansions.

The environment that was common to the faunal elements of a centre during the latest regressive phase has been a refuge area whose function is still evident today from the fact that it is identified as the place where many taxa of the most different plant and animal groups took refuge.

The function of a centre as a place of refuge is further patent from the varying degree of differentation of its faunal elements; MÜLLER (1970) says: 'Monotypic genera, species (of polytypic genera) and subspecies occur simultaneously in the same centre'.*

Even if the richness of an area in forms could be interpreted with reasons of recent ecology in such a way that the area could be considered as the place where all elements involved find an optimum of living conditions, this would not supply a satisfying explanation of the phenomenon that taxa of the most varying levels of differentiation show a congruence of their ranges in the nuclear areas defined by DE LATTIN (1957) as dispersal centres.

The fact that higher taxa such as monotypic genera are bound to a specific dispersal centre does not, however, permit any definite conclusion as to the isolation age of the centre, with the exception, perhaps, of islands. It is rather quite possible that a supraspecific taxon which is a faunal element of a dispersal centre still discernible today, has been a faunal element of another centre during previous regressive phases.

The problems involved in the necessity to differentiate between the dispersal processes of different time levels can only be countered by using systematically real units in the sense of DE LATTIN (1957), in order to obtain a basis for comparison.

It is for instance also problematic to associate monotypic genera with centres in areas with several neighbouring dispersal centres, as they occur in Central America in the area of the faunal border.

Monarda oryx (cf. fig. 17, p. 83).could be unequivocally identified as a faunal element of the Mexican and, hence, the southernmost Nearctic arboreal centre both due to its relationship with other Nearctic genera and the position and altitude of its localities, whereas it cannot so far be stated to which centre *Tro-*

* 'Nebeneinander kommen im gleichen Zentrum monotypische Genera, Spezies (von polytyischen Genera) und Subspezies vor'.

golegnum pseudambulyx belongs. This likewise monotypic genus from the Neo-tropical-Neartic border area occurs in Hidalgo, Mexico, in altitudes of 2,750 m. Its closest relative, however, is the genus *Orecta* which consists of specifically and subspecifically differentiated faunal elements of southern Neotropical centres (fig. 24, p. 106). The spatial and temporal distance between these related genera permits us, if anything, just to guess at their divergent range development.

Things are different with polytypic and polycentric species whose subspecifically differentiated populations must necessarily have an allopatric distribution for reasons of evolutionary genetics (MÜLLER 1970, 1973).

A comparison of the subspecies of a species distributed among different dispersal centres which will be the object of the following relationship analysis, provides insight into the process of the geographical speciation. This applies also to other geographically isolated taxa such as the members of a superspecies, as demonstrated by MÜLLER (1970, 1973) on the example of the *Crax rubra* superspecies complex (Aves) analysed by VUILLEUMIER (1965).

The beginning of the isolation phase of young faunal elements of a centre necessary for differentiation lies in the time in which the centre last functioned as an area of regression.

The comparative consideration of the ranges of most closely related taxa gives a clue to their subsequent dispersal. The development that took place in the dispersal centres coincided with the genesis of the respective landscapes.

In view of the fact that the history of ranges is at the same time the history of the respective taxa, a comparative study of the ranges of the related taxa is also of importance for phylogeny; MAYR (1964, p. 476) says: 'Much can be learned about the former history of taxa from consideration of the distribution of related taxa'.

b. Relationship analysis

In systematics, the speciation which the Sphingidae have undergone in the New World and in the Neotropical region respectively is conceived of as follows on the basis of recent taxa:

	Genera	Polytypic genera	Species	Polytypic species	Sub-species
New World	60	40	432	91	235
Neotropical region	45	32	383	88	223

In the New World, the genera which are especially rich in species are *Manduca* with 65 species and *Sphinx* with 36 species among the Sphingini, *Protambulyx* and *Amplypterus* with 11 species each among the Smerinthini and *Nyceryx* with 18 species among the Dilophonotini. *Eumorpha*, the only American genus of the tribe Philampelini, has 21 species. *Xylophanes* with as much as 84 species is not only the genus with the greatest number of species within the New World but also the richest of all spingid genera on a world-wide level.

Excluding *Sphinx pinastri* and *Hyles euphorbiae* which belong to the Old World, the above polytypic species consist of the following number of subspecies:

60 species with 2 subspecies
16 species with 3 subspecies
 7 species with 4 subspecies
 2 species with 5 subspecies (Xylophanes amadis, Xylophanes chiron)
 2 species with 6 subspecies (Manduca diffissa, Manduca rustica)
 1 species with 7 subspecies (Manduca sexta)
 1 species with 8 subspecies (Isognathus rimosus)

Isognathus rimosus (fig. 26) has been described to consist of 8 subspecies, of which 6 could be definitely recognized as faunal elements of the established centres.

The nominate form occurs on Cuba. Among the West Indian Islands, another 3 subspecies are known to inhabit Jamaica, Hispaniola and Puerto Rico. The subspecies living on the mainland are faunal elements of the Venezuelan centre, the Yungas centre and the Serra do Mar centre, from where the respective subspecies have experienced but a limited dispersal. In contrast to this, the subspecies *inclitus* occurs all over Central America and sporadically also in the Andes in Ecuador and Peru. It has its main distribution in the Central American rainforest centre, from where its dispersal seems to have taken place.

The remaining *Isognathus* species are monotypic; with the exception of *Isognathus tepuyensis* which inhabits the Guiana Highlands in Venezuela and *Isognathus australis* which lives in the Serra do Mar, they occur predominantly in the tropical lowland rain forest.

Manduca sexta is a very wide-spread euryecious species which occurs in the Nearctic region with its nominate form, on all West Indian Islands with the subspecies *jamaicensis* and likewise polycentrically in the Neotropical region with the ssp. *paphus*.

The remaining 4 subspecies are monocentric and represent faunal elements of the Yungas centre, the Tucuman centre, the Andean Pacific centre (fig. 25, p. 109) and the Galápagos centre (fig. 10, p. 70). The wide distribution of the three first mentioned subspecies may be connected with the fact that, among others, they feed on cultivated Solenaceae, on which they can even become a pest.

Manduca diffissa is an exclusively Neotropical species which is not even represented on the West Indian Islands and which is replaced by *Manduca occulta* in Central America. The outer appearance of the ssp. *tropicalis* does not permit a separation from *Manduca diffissa*, but the subspecies differs from the latter by its genitalia (cf. MOOSER 1940).

The ssp. *tropicalis* has its main distribution in the Andes from Peru to Venezuela, but it also occurs on Trinidad and in the Guianas and, sporadically, also in the Amazon area.

117

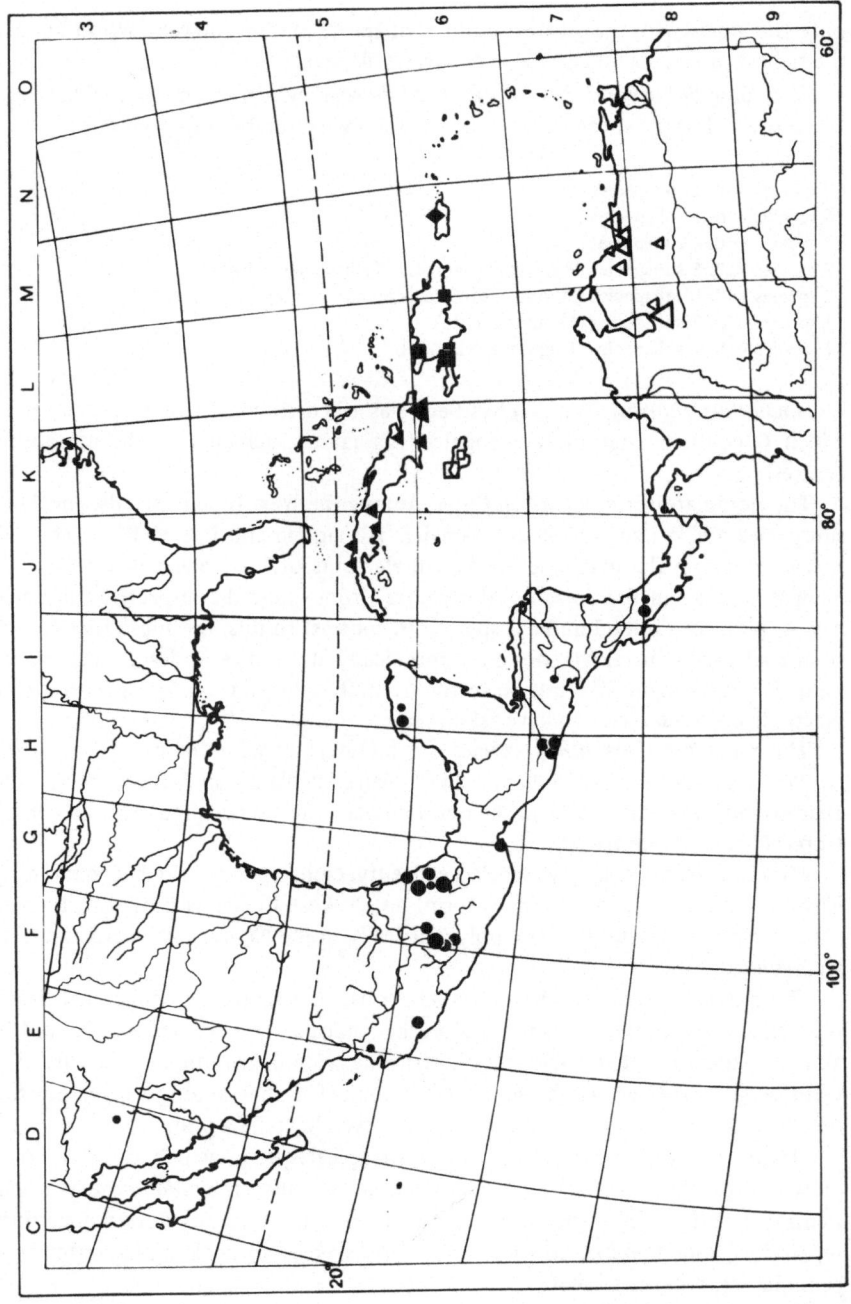

Fig. 26. Distribution of the subspecies of *Isognathus rimosus* (GRT.): *Isogn. rimosus rimosus* (GRT.) ▲, *Isogn. rimosus inclitus* EDW. ●, *Isogn. rimosus papayae* BDV. △, *Isogn. rimosus wolcotti*: CLARCK ◆, *Isogn. rimosus jamaicensis* R. & J. □.

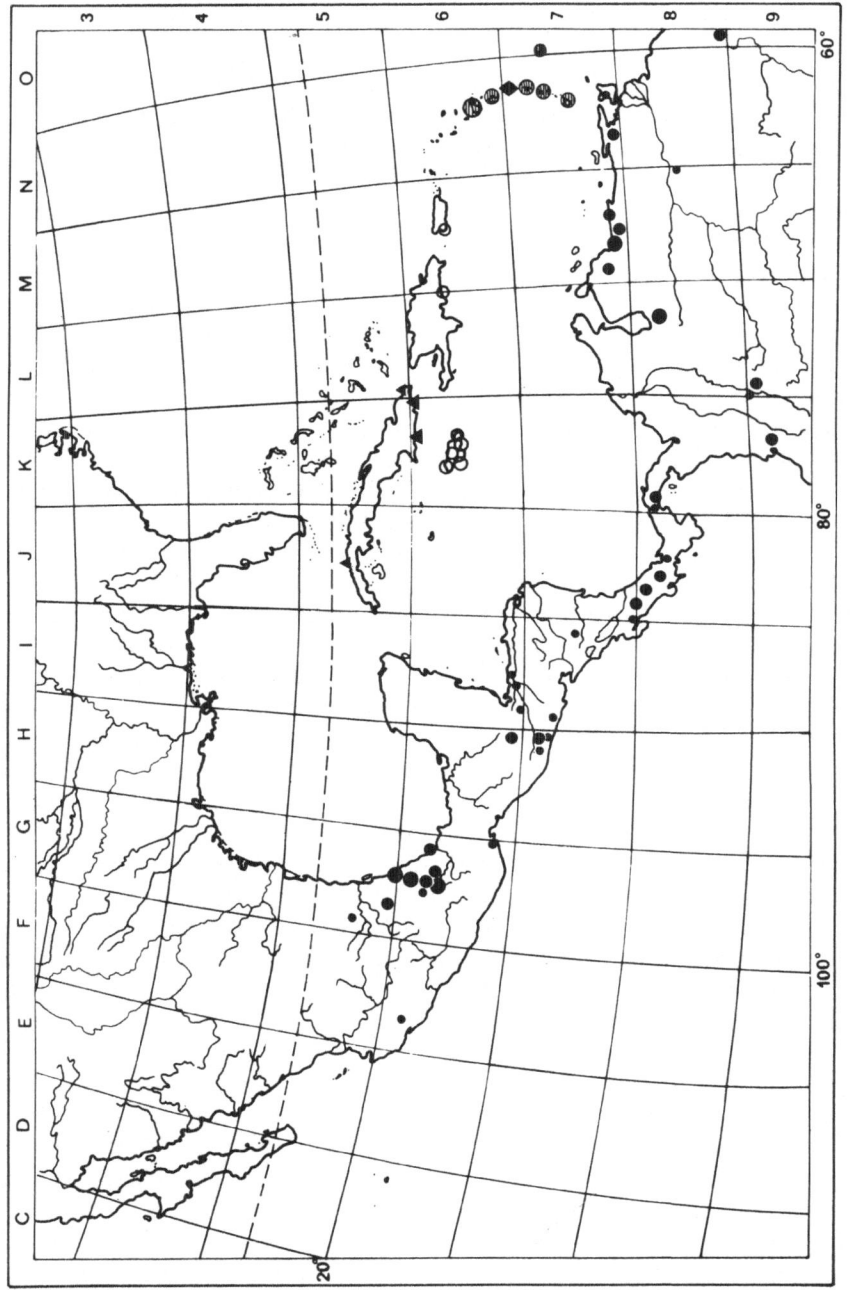

Fig. 27. Distribution of the subspecies of *Xylophanes chiron* DRURY: *Xyl. chiron chiron*
DRURY ○, *Xyl. chiron nechus* (CRAM.) ●, *Xyl. chiron martiniquensis* KERNBACH ◆, *Xyl.*
chiron lucianus R. & J.◉, *Xyl. chiron cubanus* R. & J. ▲.

119

Geographical variation of this species is restricted to the south of the distribution range in that 4 subspecies are associated with one of the following centres each: Cochabamba subcentre within the Yungas centre, Tucuman centre, Serra do Mar centre and Uruguayan centre (fig. 24, p. 106). The ssp. *petuniae* appears to have spread from the Serra do Mar centre southward and westward, which, according to the locality data recorded, might possibly have led to an early mixing with the forms from other centres. CLARK (1927) described the ssp. *ochracea* from Buenos Aires; the overlap of its range with that of the nominate form from the Uruguayan centre likewise calls for closer investigation. It would be worthwile to search for the possible existence of hybrid belts between such centres.

In contrast to *Manduca diffissa*, *Manduca rustica* is represented on the continent only by one subjspecies which occurs from the south of the United States down to Argentina. The described geographical variation is restricted to islands. Subspecies are known from the Galápagos Islands, Cuba, Hispaniola and the Lesser Antilles; they can be considered as faunal elements of the respective centres. CARY (1963) described the ssp. *cortesi* from Baja California; it differs from the nominate form mainly by its smaller size and its darker coloration.

Xylophanes chiron (fig. 27) is also composed of a wide-spread euryecious continental race, the ssp. *nechus*, and populations which are differentiated in several geographical races on the West Indian Islands. In the north, the range of the continental subspecies extends as far as the Central American rainforest centre, in the south as far as Cochabamba and South Brazil. The distribution limits seem to be connected with low temperatures, an observation which is also confirmed on the West Indian Islands. The subspecies inhabiting Cuba is restricted to the southeast of the island, where an East Cuban dispersal centre has been defined. The nominate form is frequent on Jamaica; a few records are known from Hispaniola and Puerto Rico. The Lesser Antilles are inhabited by the ssp. *lucianus*, however the specimens recorded from Guadeloupe might belong to the nominate form. KERNBACH (1964) described a distinct subspecies from Martinique.

The distribution pattern of *Xylophanes chiron* on the Antilles can be compared with the results of studies by CLENCH (1963) on West Indian Lycaenidae (fig. 28). With the exception of the extreme eastern part of Cuba, CLENCH excluded the largest part of this island as well as Puerto Rico and the Virgin Islands as possible refuge areas of the Lycaenidae during the latest glaciation period on the grounds that these areas have been exposed to lower temperatures than the remaining West Indian Islands. This would explain the fact that no dispersal centre for sphingids could be established on Puerto Rico. Furthermore, it would explain the position of the East Cuban centre historically and argue against the exclusive responsibility of reasons of recent ecology, according to which one might tend to maintain that the mountains in the east of Cuba are the least influenced anthropogenicly and, hence, particularly rich in species.

Xylophanes amadis is missing on the West Indian Islands; on the continent, it is predominantly distributed in the north of the Neotropical region (fig. 29). Only the ssp. *cyrene* is polycentrically and almost disjunctly distributed in the Costa

120

Rican centre and the Central American rainforest centre. All the other subspecies are faunal elements of described centres, i.e. the Yungas centre, the Venezuelan centre and the Guiana centre; within the latter, the nominate form occurs in French Guiana and the ssp. *goeldii* in Pará, a fact that requires a subdivision into two secondary centres.

Just like the explained examples, the subspecies of all the remaining polytypic species can be prima facie understood as faunal elements of the described centres, provided that they are monocentric; as such they have already been mentioned in the description of the different centres together with the monotypic species with small ranges. Furthermore, polycentric species and subspecies can be associated with given centres, if they show a concentration of occurrence within such centres, or if their range patterns show a definite spreading from these centres.

Another possibility of judgement flows from the exact knowledge of the eco-logical valency of the taxa, on the one hand, and the knowledge of the ecological conditions prevailing in the differentiation centres, on the other.

On the assumption that the ecological valency is relatively stable (cf. HEYDE-MANN 1943) closely related taxa can be expected to show relations with centres which, for their part, are characterized by similar ecological conditions.

Setting out from these considerations, DE LATTIN (1957) devided the dis-

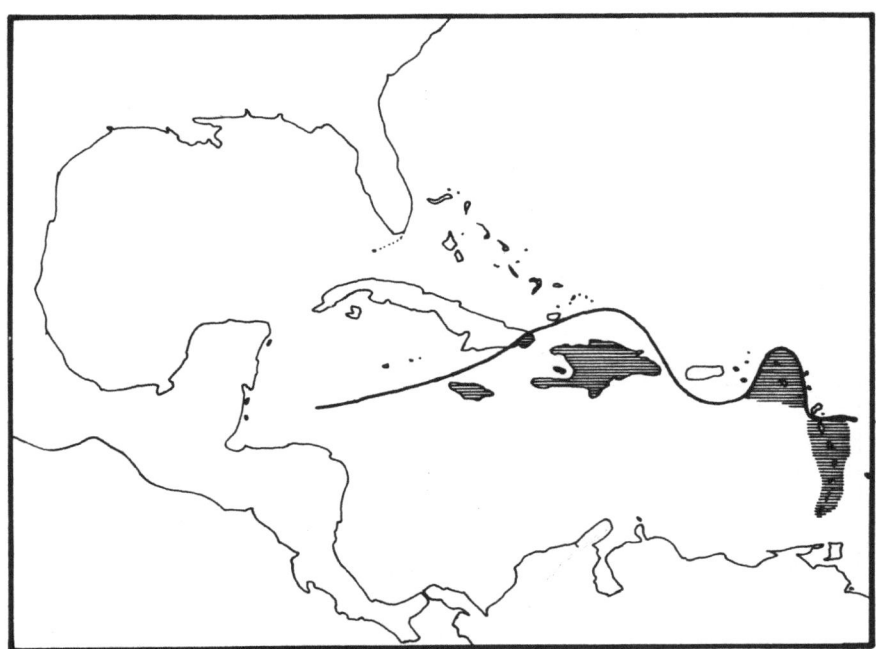

Fig. 28. Presumed refugia during Wisconsin Glacial; line representing approximate position of 66°F isotherm (from CLENCH 1963).

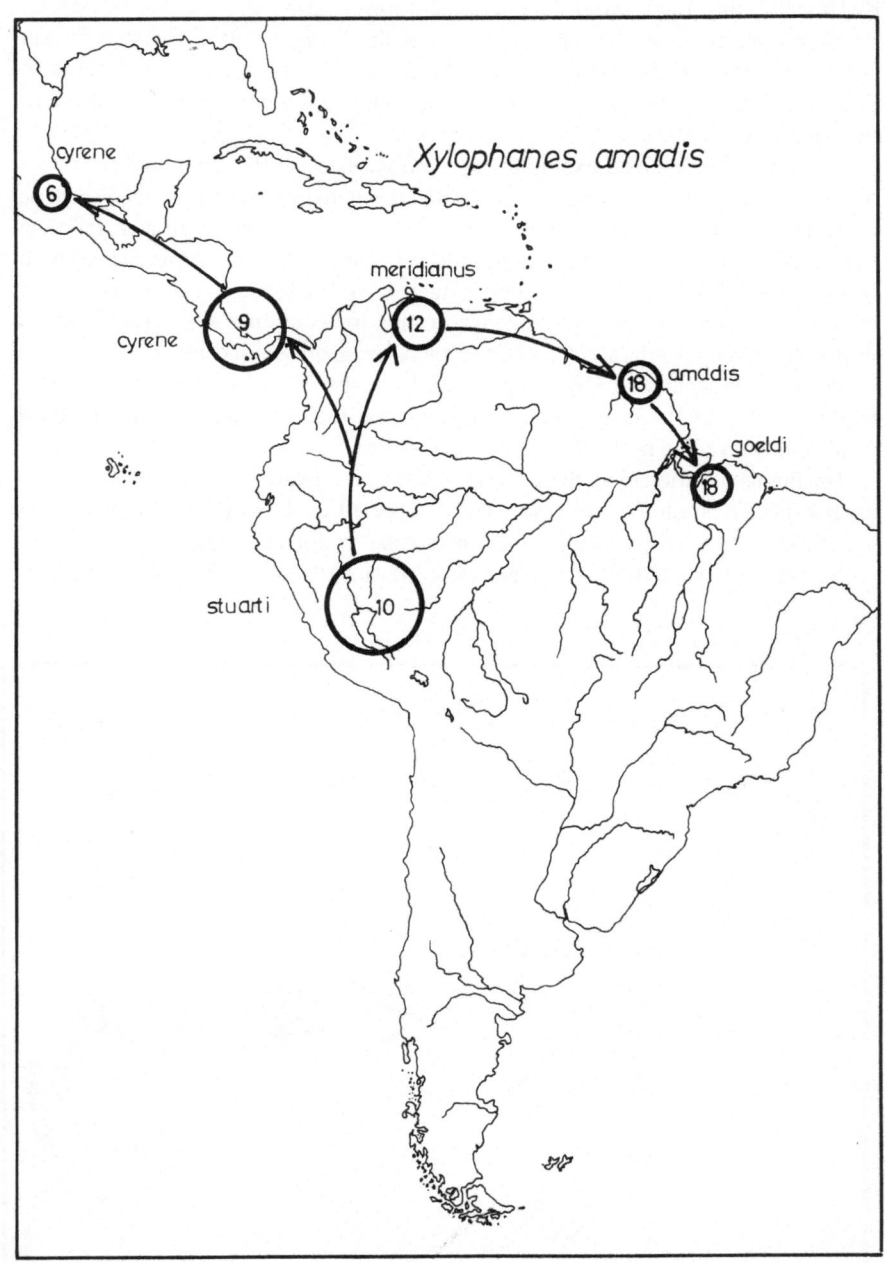

Fig. 29. Distribution and assumed dispersal of *Xylophanes amadis* (STOLL); numbers in circles correspond to numbering of dispersal centres (fig. 8).

persal centres of the Holarctic terrestrial fauna into centres of the Arboreal, the Eremial and the Boreal according to the three major macro-ecological environments as defined by REINIG (1937). In contrast to DE LATTIN (1957) who did not think that any further ecological subdivision of the centres was possible, MÜLLER (1970, 1971, 1973) arranged the arboreal centres and the centres of non-forest areas analysed by him in affinity groups according to their degree of relationship. In accordance with the vegetation formations existing in the Tropics, the Arboreal was identified with the rain-forest and, hence, understood in a much narrower sense than by DE LATTIN (1957) who conceived of the Arboreal as an ecologically more complex environment reaching 'from closed forest to "Steppenheide", from flat moor to karst formation, from the taiga to macchia'.*

The narrower conception of the Arboreal automatically entails a broadening of the concept of Eremial (cf. MÜLLER 1970, 1973). By Eremial DE LATTIN (1967) only understood the extreme arid zones of the earth including genuine steppes, arid steppes and deserts, even though he confesses with regard to the definition of the Eremial and the Arboreal in the Tropics (p. 258): 'In the Tropics, the definition of the Eremial as against that of the Arboreal causes greater difficulties, because, even though it is true that the distinct arid forest and arid savannas of the Tropics are regions characterized by an extremely dry and hot climate, they are nevertheless more or less densely overgrown by specific trees and bushes adapted to such extreme conditions'.**

In view of the marking off of all rain-forest-free areas below 1,500 m from the tropical Arboreal (MÜLLER 1970, 1971, 1973) it seems more correct to avoid the term of the Eremial altogether.

Due to the fact that the sphingids cannot survive in areas with extremely low temperatures, oreal centres are completely missing from the forest-free environments of the high mountains above the tree-limit.

The greatest density of sphingid species and individuals is to be found (SEITZ 1931, p. 839) 'in the humid forest-districts of tropical and subtropical countries'***; 'in the vegetation-free deserts' their number 'occasionally drops to zero'****. This general distribution tendency is expressed and confirmed by the arboreal position of the centres.

The group of the following centres is to be considered as arboreal:

2, 3, 4, 5, 6, 9, 10, 11, 12, 13, 16, 18.

---

* 'vom geschlossenen Wald bis zur Steppenheide, vom Flachmoor bis zur Karstformation, von der Taiga bis zur Macchie'
** 'In den Tropen stößt die Abgrenzung von Eremial und Arboreal auf größere Schwierigkeiten, weil wir hier in den ausgesprochenen Trockenwäldern und in den Trockensavannen Gebiete vor uns haben, die zwar durch ein außerordentlich trockenes und heißes Klima ausgezeichnet sind, die aber nichtsdestoweniger einen mehr oder weniger dichten Bewuchs von an derart extreme Verhältnisse angepaßten Bäumen und Sträuchern tragen'.
*** 'in den wasserreichen Walddistrikten tropischer und subtropischer Länder'
**** 'sinkt in den vegetationslosen Wüsten stellenweise auf Null'.

123

The arboreal centres are mutually correlated by close relationships of their faunal elements. They can be set over against the following centres located in forest-free areas below 1,500 m:

1, 7, 8, 14, 15, 17.

These centres, however, are not as a group characterized by the relationship of their faunal elements, which either points to their heterogeneousness or is an expression of their isolated position.

The following illustration shows the relations between centres as they result from the relationship between two subspecies each of a sphingid species that have been defined as faunal elements of different centres.

	1	2	3	4	5	6	7	8	9	10	11	12	13	14	15	16	17	18
1	0	1	1	2	1	0	0	0	0	1	0	0	0	1	1	0	0	0
2		0	4	5	4	1	1	0	0	1	0	1	1	0	0	0	0	0
3			0	2	1	1	0	0	0	1	0	1	1	0	0	0	0	0
4				0	3	0	1	0	0	1	0	1	1	0	0	0	0	0
5					1	0	0	0	0	0	0	0	0	0	0	0	0	0
6						0	0	0	1	1	1	2	2	0	0	0	0	0
7							0	0	1	0	0	0	0	0	0	0	0	0
8								0	0	0	0	0	0	0	0	0	0	0
9									0	1	1	0	1	1	0	0	0	0
10										1	2	3	6	1	1	1	0	2
11											0	1	1	2	1	0	0	0
12												0	1	0	0	0	0	2
13													0	3	0	0	0	0
14														0	0	0	0	0
15															0	0	0	0
16																0	0	0
17																	0	0
18																		1

The rate of recurrence of combinations between the same centres or subcentres can be considered as a criterion for the natural affinity of these centres (see p. 125).

An inclusion of most closely related species in this comparison would further substantiate the affinity of the centres. Through a 'phylogeny of the dispersal centres' one could even penetrate as far as to the centres of origin of higher taxa.

In grouping ecologically related centres which are not connected with each other in terms of their position, an affinity of the centres can also be recognized on the basis of mutual connections existing between them and a third centre.

The relationship between the centres 2, 3, 4 and 5 within the West Indian Islands can be considered as natural both on the basis of comparable ecological conditions and as an expression of their geographical neighbourhood.

The very high degree of affinity between the Yungas centre and the Serra do Mar centre does not only manifest itself by subspecies pairs which both centres

6x	5x	4x	3x	2x	1x
				1+4	1+2, 3, 5, 10, 14, 15
	2+4	2+3, 5			2+6, 7, 10, 12, 13
				3+4	3+5, 6, 10, 12, 13
			4+5		4+7, 10, 12, 13
					5+5
				6+12, 13	6+9, 10, 11
					7+9
					9+10, 11, 13, 14
10+13			10+12	10+11, 18	10+10, 14, 15, 16
				11+14	11+12, 13, 15
				12+18	12+13
			13+14		
					18+18

have in common. Many species have a disjunct distribution in the Andes on the one hand and the Serra do Mar on the other (fig. 62, p. 170). Other species occur with one subspecies polycentrically in the Andes and in Central America and with another subspecies in the Serra do Mar (fig. 63, p. 171). This centre affinity finally expresses itself by closely related species for which a sibling-group relationship can be assumed on the basis of the separation between the two mutually related areas of differentiation.

The search for sibling groups in the sense of HENNIG (1969) can be carried through parallelly in the comparative morphological and chorological way by investigating species and subspecies in most closely related centres.

The fact that the ecologically different centres 13 and 14 (Serra do Mar centre and Uruguayan centre) share different subspecies each of three species (fig. 24, p. 106) could be explained by vegetation fluctuations in this area. Differentiation could have occurred subsequent to the penetration of forest species into the river-bank forests of the open landscapes in the course of a post-glacial southern expansion of the coastal rain-forest. During an arid phase which is known to have taken place from 6000 to 2400 B.C. (MÜLLER 1970, 1973) the foremost populations were isolated and could differentiate subspecifically.

The process of speciation between two forests centres could be similarly conceived of, although it need not be supposed that two rain-forest areas which are separated today had been directly connected with each other during climatically more humid phases. What matters is that during those periods expansive species succeeded in transmigrating shrunk arid areas which could then develop their efficacity to the full in the course of a subsequent dry phase.

At the end of the post-glacial aridity phase an expansion of the rain-forest set in (MÜLLER & SCHMITHÜSEN 1970) which, although in many places seemingly regressive under man's influence, has continued to this day and resulted in the spreading of forms adapted to the forest.

In the tropical lowland, speciation can be explained by alternating arid and humid phases and associated vegetation fluctuations, a phenomenon which has been particularly well explained by HAFFER (1969).

As for montane forest biomes to which a great variety of sphingids are adapted within the Neotropical region (cf. fig. 45, p. 153), vertical biochore shifts have to be assumed which occurred in correlation with the Quartenary cold periods on the north continents (cf. VUILLOMIER, B.S. 1971, WILHELMY 1952). This could also have led to range developments which enabled geographical speciation to take place wherever different populations of a species have been isolated from each other sufficiently long.

With the elimination of the isolation conditions a more or less rapid expansion of the isolates took place depending on their locomotion capacity, their ecological valency, the population density and their genetic constitution, and in the course of this process the limits of the differentiation centres were bound to vanish more and more.

However, as evidenced by the distribution analysis of Neotropical sphingids, these centres can be reconstructed on the basis of the ranges of static elements and identified as the nuclear areas of the ranges of expansive types.

The sometimes complex recent distribution pattern and the richness of Neotropical biomes in forms can be understood as a result of the described range dynamics; in contrast to the north continents, however, geographical speciation in the Neotropical region could pass off without any extensive extinction.

Modern biogeographical research endeavours to make statements about the genesis of ranges and landscapes and to gain knowledge about the speciation and phylogeny of the investigated organisms, not speculatively, but by means of a comparative chorological interpretation of the available data material.

The present work is based on the method developed by DE LATTIN (1957) by which he analysed the dispersal centres of the terrestrial fauna of the Holarctic region.

The universality of the concept of dispersal centres and its applicability to the Tropics have been proved by MÜLLER (1970, 1973) who analysed a total of 40 Neotropical centres on the basis of the interpretation of vertebrate ranges. The purpose of the present work was to revise and, if possible, confirm the results of MÜLLER's earlier investigations by a chorological study of the Neotropical Sphingidae.

In applying this method it is of paramount importance to consistently restrict the studies to the ranges of species and subspecies which alone can be considered as real units. It is a further prerequisite that the groups to be investigated are sufficiently clarified both with regard to their taxonomic and their chorological status.

The Sphingidae appeared to fulfil these conditions best within the Neotropical invertebrates. By examining the sphingid collections of 17 European and North American museums more than 74,000 locality data entries could be registered and recorded in range maps. Furthermore, light trap collecting has been done in the area of the faunal border between the Nearctic and the Neotropical regions, an area discussed in connection with the delimitation of the area under investigation.

In view of the fact that only part of the total of 576 drawn range maps could be included in this work as sample-maps of the established centres, a tabular survey of the distribution of all sphingid species and subspecies of the New World was added, after their systematics had been checked on the basis of recent revisions. In addition, a discription of the family Sphingidae was given, and a series of questions as to their nomenclature was discussed.

As shown from a numerical recording of the species in the units of a grid-map based on geographical longitudes and latitudes, the sphingids are by no means distributed evenly throughout the Neotropical region. From this survey it appears most clearly that a richness in taxa exists in the areas which could be defined as distribution and dispersal centres in the course of the subsequent investigations.

A total of 18 centres could be shown to exist in the area covering Central and

South America including the West Indian and the Galápagos Islands; the majority of these centres proved to be congruent with the centres analysed by MÜLLER (1970, 1973).

The rain-forest centres of the montane and submontane altitudes could be verified best. Within the Hylaea, on the other hand, no centre could be established; apart from a lack of sufficient locality data, the reason for this fact must be seen in missing or eliminated isolation conditions. In contrast to this, ideal isolation conditions are found on islands even recently, as can be seen from the high share the Galápagos Islands have in sphingid endemics.

In a relationship analysis of the centres, a group of arboreal centres could be set over against the centres of forest-free areas. In order to find out which centres show the highest degree of relationship with each other, the rate of occurrence of a subspecific differentiation of polytypic species in two centres each has been examined.

In view of the fact that geographical speciation presupposes conditions enabling a temporary isolation of the populations, the range development during the most recent differentiation phase between related centres can best be explained with the assumption of Quarternary vegetation fluctuations.

The richness of Neotropical biomes in taxa can be understood as a result of geographical speciation without any extensive extinction.

VIII. BIBLIOGRAPHY

ALLEN, J.A. (1871): On the mammals and winter birds of East Florida. Bull. Mus. Com. Zool. 3: 161-450.

AMSEL, H.G. (1938): *Amphimoeca walkeri* BSD., der Schwärmer mit dem längsten Rüssel. Ent. Rundschau 55 (15): 165-167.

ANT, H. (1964): Der boreoalpine Verbreitungstyp bei europäischen Landgastropoden. Verhdl. Dtsch. Zool. Ges. Kiel.

BARNES, W. & A.W. LINDSAY (1922): A review of some generic names in the order Lepidoptera. Ann. Ent. Soc. Amer. 15 (I): 89-99.

BARRERA, A. (1962): La peninsula de Yucatán como provincia biótica. Rev. Soc. Mex. Hist. Nat. 23.

BARTHOLOMEW, J.G., CLARKE W.E. & P.H. GRIMSHAW (1911): Atlas of Zoogeography. In: Bartholomew's Physical Atlas 5, London.

BATES, M. (1935): The butterflies of Cuba. In: Bull. Mus. Com. Zool. Vol. LXXVIII, no. 2: 65-258.

BAUMANN, H. & E. REISSINGER (1969): Zur Tagfalterfauna des Chanchamayogebietes in Peru. I. Einleitung. Pieridae. Veröff. Zool. Staatssamml. München. 13: 71-142.

BEEBE, W. (1924): Galápagos: World's end. G.P. Putnam's Sons, New York.

BEEBE, W. & H. FLEMING (1945): The Sphingidae (Moths) of Kartabo, British Guiana, and Caripito, Venezuela. In: Zoologica. New York Zool. Soc. 30 (1): 1-6.

BIEZANKO, C.M. DE & J. ZOPP (1954): Die Sphingiden Uruguays. Ent. Z. 64 (11/12/13).

BLYTH, E. (1871): A suggested new division of the earth into zoological regions. Nature 3: 427.

BOWMAN, R.I. (1966): The Galápagos. Proceedings of the Galápagos International Scientific Project of 1964. University of California Press, Berkeley and Los Angeles.

BROWN, K.S.Jr. (1975): Geographical patterns of evolution in Neotropical Lepidoptera. Systematics and derivation of known and new Heliconiini (Nymphalidae: Nymphalinae). J. Ent. (B) 44 (3): 201-242.

BURMEISTER, H. (1856): Systematische Übersicht der Sphingidae Brasiliens. Abhandl. Naturf. Ges. Halle, 3, Sitzungsberichte: 58-74.

CABRERA, A. & J. YEPES, (1940): Mamiferos sudamericanos. Cia. Argent. Edit., Buenos Aires.

CABRERA, A. & J. YEPES, (1947): Zoogeografia. In: Geografia de la Republica Argentina. Gaea 8: 347-483.

CANDOLLE, A.P. DE (1855): Géographie botanique raisonnée. 2 vols. Masson, Paris and Geneva.

CARCASSON, R.H. (1968a): The Sphingidae (Hawk Moths) of Eastern Africa. Thesis. University of East Africa.

CARCASSON, R.H. (1968b): Revised catalogue of the African Sphingidae (Lepidoptera) with descriptions of the East African species. Jour. East Africa Nat. Hist. Soc. and Natl. Mus., 26 (3): 1-148.

CARY, M.M. (1949): Sphingidae collecting in North-Central Venezuela in June, 1949. The Lepidopterists' News (New Haven) 3 (7): 78.

CARY, M.M. (1951): Distribution of Sphingidae (Lepidoptera Heterocera) in the Antillean-

129

Caribbean Region. In: Trans. American Ent. Soc., 77: 63-129.

CARY, M.M. (1963): Reports on the Margaret M. CARY and Carnegie Museum expedition to Baja California, Mexico 1961. 2. The family Sphingidae (Lepidoptera). Ann. Carnegie Mus, 36: 193-204.

CLARK, B.P. (1917): New Sphingidae. Proc. New England Zoological Club, 4: 57-72.

CLARK, B.P. (1919): Some undescribed Sphingids. Proc. New England Zoological Club, 6: 99-114.

CLARK, B.P. (1922): Twenty-five new Sphingidae. Proc. New England Zoological Club, 8: 1-23.

CLARK, B.P. (1923): Thirty-three new Sphingidae. Proc. New England Zoological Club, 8: 47-77.

CLARK, B.P. (1926): A Revision of the Protoparces of the Galápagos Islands. Proc. New England Zoological Club, 9: 67-71.

CLARK, B.P. (1927): Descriptions of 12 new Sphingids and remarks upon some other species. Proc. New England Zoological Club, 9: 99-104.

CLARK, B.P. (1931): Descriptions of seven new Sphingidae and a note on one other. Proc. New England Zoological Club, 12: 77-83.

CLENCH, H.K. (1963): A synopsis of the West Indian Lycaenidae with remarks on their zoogeography. In: Journal of Research on the Lepidoptera 2 (4): 247-270.

CURIO, E. (1965): Die Schutzanpassungen dreier Raupen eines Schwärmers (Lepidoptera, Sphingidae) auf Galápagos. Ergebnisse der Deutschen Galápagos-Expedition 1962/63. In: Zool. Jb. Syst. 92: 487-522.

DANIEL, F. (1949b): Liste der von Pater Cornelius Vogl in Maracay und Caracas gesammelten XXXV-XXXIX: 230-234.

DANIEL, F. (1949b) Liste der von Pater Cornelius Vogl in Maracay und Caracas gesammelten Schmetterlinge. II. Sphingidae. Bol. Entom. Venez. (Caracas) 8 (1-2): 21-42.

DARLINGTON, P.J.Jr. (1938): The origin of the Greater Antilles with discussion of dispersal of animals over water and through the air. Quart. Rev. Biol. 13: 274-300.

DARLINGTON, P.J.Jr. (1957): Zoogeography: The geographical distribution of animals. John Wiley and Sons, New York.

DARWIN, C. (1859): On the origin of species by means of natural selection, or the preservation of favoured races in the struggle for life. John Murray, London.

DARWIN, C. (1862): Fertilisation of orchids. Dtsch. Übersetzung von J.V. Carus: Die verschiedenen Einrichtungen, durch welche Orchideen von Insekten befruchtet werden. Stuttgart 1877.

DAVIDSON, T. (1965): Moths that behave like hummingbirds. In: National Geographic, 127 (6).

DENBURGH, J. VAN & J.R. SLEVIN (1913): Expedition of the California Academy of Sciences to the Galápagos Islands, 1905-1906. Proc. Calif. Acad. Sciences 2 (1): 133-202.

DIERL, W. (1970): Grundzüge einer ökologischen Tiergeographie der Schwärmer Ostnepals (Lepidoptera: Sphingidae). Khumbu Himal 3 (3): 313-360.

DRAUDT, M. (1931): Familie Sphingidae. In: A. Seitz: Die Gross-Schmetterlinge der Erde. 6: 845-900, pl. 90-98. Alfred Kernen Verlag, Stuttgart.

EDITÔRA GLOBO (1967): Dicionário Geográfico Brasileiro. Editôra Globo S.A., Pôrto Alegre.

EIBL-EIBESFELDT, I. (1959): Survey on the Galápagos Islands, Paris.

EIBL-EIBESFELDT, I. (1960): Galápagos, die Arche Noah im Pazifik. München.

ENGLER, A. (1882): Versuch einer Entwicklungsgeschichte der Pflanzenwelt, insbesondere der Florengebiete seit der Tertiärperiode. II. Die extratropischen Gebiete der südlichen Hemisphäre und die tropischen Gebiete. Engelmann, Leipzig.

ERWIN, T.L. (1970): A reclassification of bombadier beetles and a taxonomic revision of the North and Middle American species (Carabidae: Brachinida). Quaest. ent. 6: 4-215.

FITTKAU, E.J. (1969): The fauna of South America. In: Biogeography and Ecology in South America 2: 624-658.

FLEMING, H. (1947): Sphingidae (Moths) of Rancho Grande, North Central Venezuela. Zoologica (New York) 23 (3): 133-145.

FLETCHER, D.S. (1966): Some changes in the nomenclature of British Lepidoptera. Part I. In: Entomologist's Gazette, 17 (1).

FORBES, W.T.M. (1930): Insects of Porto Rico and the Virgin Islands. Heterocera or moths (excepting the Noctuidae, Geometridae and Pyralidae). Scientific survey of Porto Rico and the Virgin Islands, 12 (1): 47-70.

FORSTER, W. (1954): Biologie der Schmetterlinge. In: Forster W. und T.A. Wohlfahrt: Die Schmetterlinge Mitteleuropas, I. Franckh'sche Verlagshandlung, Stuttgart.

FORSTER, W. (1958): Die tiergeographischen Verhältnisse Boliviens. Proc. X. Int. Congress of Entom. I. 843-846.

FORSTER, W. & T.A. WOHLFAHRT (1956): Die Schmetterlinge Mitteleuropas. III, Spinner und Schwärmer. Franckh'sche Verlagshandlung, Stuttgart.

FOX, J.W. (1962): Reports on the Margaret M. Cary and Carnegie Museum expedition to Baja California, Mexico 1961. 3. A portable ultra-violet insect trap. Ann. Carn. Mus. 36: 205-212.

FREITAG, R. (1969): A revision of the species of the genus *Evarthrus* LE CONTE (Coleoptera: Carabidae). Quaest. ent. 5: 89-212.

GEHLEN, B. (1935): Sphingidae aus Venezuela. Veröff. dtsch. kol. Mus. Bremen 1 : 305-306.

GEIYSKES, D.C. (1934): Notes on the Odonate fauna of the Dutch West Indian Islands Aruba, Curacao and Bonaire. Int. Rev. Hydrobiol. 31: 287-311.

GOLDMAN, E.A. & R.T. MOORE (1946): The biotic provinces of Mexico. Journ. of Mammalogy 26 (4): 347-360.

GOOD, R. (1964): The Geography of flowering plants. Longmans, London.

GROSS, F.J. (1961): Zur Geschichte und Verbreitung der euro-asiatischen Satyriden (Lepidoptera). Verh. Dtsch. Zool. Ges. Bonn (1960).

GROSS, F.J. (1962): Zur Evolution euro-asiatischer Lepidopteren. Verh. Dtsch. Zool. Ges. Saarbrücken (1961).

GROTE, A.R. (1865): Notes on Cuban Sphingidae. Proc. Ent. Soc. Philadelphia 5: 38-84.

GROTE, A.R. & C.T. ROBINSON (1865): A synonomical catalogue of North American Sphingidae with notes and descriptions. Proc. Ent. Soc. Philadelphia 5: 149-193.

GUNDLACH, J. (1881): Contribución a la Entomologia Cubana. Habana.

HAFFER, J. (1967): Speciation in Colombian forest birds west of the Andes. Amer. Mus. Novitates 2294: 1-57.

HAFFER, J. (1969): Speciation in Amazonian forest birds. Science 165: 131-137.

HAFFER, J. (1970): Geologic-climatic history and zoogeographic significance of the Uraba region in Northwestern Colombia. Caldasia 10 (50): 603-636.

HAYES, A.H. (1975): The larger moths of the Galápagos Islands (Geometroidea: Sphingoidea & Noctuoidea). Proc. Calif. Acad. Sciences XL (7): 145-208.

HEATH, J. & J. LECLERCQ (1970): Erfassung der europäischen Wirbellosen. Ent. Z. 80 (19): 195-196.

HEMMING, F. (1937): Hübner. A bibliographical and systematic account of the entomological works of Jacob Hübner and of the supplements thereto by Carl Geyer, Gottfried Franz von Fröhlich and Gottfried August Wilhelm Herrich-Schäffer. Ent. Soc., London.

HENNIG, W. (1950): Grundzüge einer Theorie der Phylogenetischen Systematik. Deutscher Zentralverlag, Berlin.

HENNIG, W. (1966): Phylogenetic systematics. University of Illinois Press, Urbana.

HENNIG, W. (1969): Die Stammesgeschichte der Insekten. Frankfurt.

HEYDEMANN, F. (1943): Die Bedeutung der 'ökologischen Valenz'. In: Ent. Z. 57 (1): 1-8.

HODGES, R.W. (1971): Sphingoidea. In: The moths of America north of Mexico. Fascicle 21. E.W. Classey Ltd. and R.B.D. Publ. Inc., London.

HOFFMANN, C.C. (1942): Catalogo sistematico y zoogeografico de los Lepidopteros Mexicanos, tercera parte: Sphingoidea y Saturnioidea. Anales del Inst. de Biologia, Mexico, 13: 213-256.

HOFFMANN, F. (1934): Beiträge zur Lepidopterenfauna von St. Catharina (Südbrasilien). Ent. Rundschau 51: 265-268, 272-277.

HOFMANN, E. (1873): Isoporien der europäischen Tagfalter. Inauguraldissertation, Jena.

HOLLAND, W.J. (1903): The moth book. Doubleday, Page & Co., New York.

HUECK, K. (1966): Die Wälder Südamerikas. Fischer Verlag, Stuttgart.

HUECK, K. & P. SEIBERT (1972): Vegetationskarte von Südamerika. Fischer Verlag, Stuttgart.

HUXLEY, T.H. (1868): On the classification and distribution of the Alectoromorphae and Heteromorphae. In: Proc. Zool. Soc. London: 294-319.

JORDAN, K. (1911): Sphingidae. In: A. Seitz: Die Gross-Schmetterlinge der Erde. 2: 229-273, Alfred Kernen Verlag, Stuttgart.

JORDAN, K. (1940): Results of the Oxford University biological expedition to the Cayman Islands, 1938. Sphingidae (Lep.). Ent. Monthly Magazine 76: 275-277.

KAESTNER, A. (1973): Lehrbuch der Speziellen Zoologie. I. Wirbellose 3. Teil. Insecta: B. Spezieller Teil. Fischer Verlag, Stuttgart.

KAYE, W.J. & N. LAMONT (1927): A catalogue of Trinidad Lepidoptera Heterocera (moths). In: Memoirs of Dept. of Agriculture, Trinidad and Tobago. 3: 80-94.

KEPKA, O. (1969): Zur Tiergeographie der Trombiculidae im Mittelmeerraum. Verh. Dtsch. Zool. Ges. 32: 526-535.

KERNBACH, K. (1952): Eine neue Unterart von *Protoparce diffissa* Btlr. aus Bolivien (Sphingidae). Z. Lepidopt. 2 (2): 129-130.

KERNBACH, K. (1956): Über *Protoparce pellenia* H.-Sch. und ihr ähnelnde südamerikanische Sphingiden. Ent. Z. 66 (5/6/7): 62, 65, 84.

KERNBACH, K. (1957): Der Ekelgeschmack der Raupe von *Celerio euphorbiae* L. Mittl. Dtsch. Ent. Ges. 16 (3/4): 51-52.

KERNBACH, K. (1960): Zur Schreckstellung der Schwärmer (Lep. Sphingidae). Ent. Z. 70 (20/21): 229-249.

KERNBACH, K. (1962a): Schwärmer mit kurzem Rüssel (Lep. Sphingidae). Dtsch. Ent. Z. N.F. 9 (III/IV): 297-303.

KERNBACH, K. (1962b): Die Schwärmer einiger Galápagos-Inseln. (Lep. Sphingidae). Opuscula Zoologica 63: 1-19.

KERNBACH, K. (1964a): Die Schwärmer einiger Galápagos-Inseln (II) (Lep. Sphingidae). Mittl. Dtsch. Ent. Ges. 23 (5/6): 88.

KERNBACH, K. (1964b): Une nouvelle sous-espèce de Sphingide de l'île de la Martinique. Bull. Soc. Ent. Mulhouse: 90-91.

KERNBACH, K. (1964c): *Protoparce sexta saliensis* ssp. n., eine neue Schwärmer-Unterart aus Südamerika (Lep. Sphingidae). Mittl. Dtsch. Ent. Ges. 23 (5/6): 89-90.

KERNBACH, K. (1965a): Ergänzungen zur Kenntnis einiger Schwärmer (I) (Lep. Sphingidae). Mittl. Dtsch. Ent. Ges. 24 (2): 86-90.

KERNBACH, K. (1965b): *Protoparce empusa* sp.n., eine neue Schwärmerart aus Venezuela (Lep. Sphingidae). Mittl. Dtsch. Ent. Ges. 24 (1): 18-20.

KERNBACH, K. (1967): Die Sphingidengattung *Sphinx* LINNE (Lep. Sphingidae) Dtsch. Ent. Z. N.F. 16 (I/III): 91-114.

KERNBACH, K. (1968): Eine neue südamerikanische Schwärmerart *Protoparce feronia* sp.n. (Lep. Sphingidae). Mittl. Dtsch. Ent. Ges. 27 (1): 5-8.

KIRBY, W.F. (1892): A synonymic catalogue of Lepidoptera Heterocera (moths). I (Sphinges and Bombyces), London.

KNAPP, R. (1965): Die Vegetation von Nord- und Mittelamerika und der Hawaii-Inseln. Fischer Verlag, Stuttgart.

KRAUS, O. (1955): Taxonomische und tiergeographische Studien an Myriapoden und Araneen aus Zentralamerika. Dissertation, Frankfurt.

KRAUS, O. (1960): Zur Zoogeographie von Zentral-Amerika (Studien an Myriapoden und Arachniden). Verh. XI. Int. Kongr. Ent. Wien I: 516-518.

KRÜGER, E. (1932): Verbreitung und Ableitung einiger Tagfalterfamilien des tropischen Amerikas (Rhop. Lep.). Dtsch. Ent. Z. 3: 149-194.

KÜHLHORN, F. (1959): Forschungen im südlichen Mato Grosso I-III. Petermanns Geogr. Mittl..

LATTIN, G. DE (1957): Die Ausbreitungszentren der holarktischen Landtierwelt. Verh. Dtsch. Zool. Ges. Hamburg 1956: 360-410.

LATTIN, G. DE (1967): Grundriss der Zoogeographie. Fischer Verlag, Jena.

LICHY, R. (1943-1948): Documents pour servir à l'étude des Sphingidae du Venezuela (Lépid. Hétér.). Bol. Ent. Venez. (Caracas) II (1): 1-16; II (2): 103-106; II (3): 157-160; III (1): 57-63; III (3): 119-124; III (4): 195-202; IV (3): 135-148; V (1): 15-26; VII (3-4): 67-89.

LICHY, R,. (1962): Documentos para servir al estudio de los Sphingidae de Venezuela (Lepidoptera, Heterocera). Rev. Fac. Agron. (Maracay) 2 (4): 53-178.

LICHY, R. (1966): Documentos para servir al estudio de los Sphingidae de Venezuela (Lep. Heterocera). Ann. Soc. Ent. Fr. (N.S.) II (2): 437-448.

LICHY, R. (1968): Documentos para servir al estudio de los Sphingidae de Venezuela (Lepidoptera, Heterocera). Bol. Acad. Cienc. Fisicas Matemáticas y Naturales 81 (18): 31-42.

LINELL, M.L. (1899): On the coleopterous insects of the Galápagos Islands. U.S. Nat. Mus. Proc. 21: 239-268.

LINSLEY, E.G. & R.L. USINGER (1966): Insects of the Galápagos Islands. Proc. Calif. Acad. Sciences 33 (7): 113-196.

LYDEKKER, R. (1896): A geographical history of mammals. Cambridge Geogr. Series, Cambridge Univ. Press.

MAACK, R. (1969): Die Serra do Mar im Staate Paraná. In: Die Erde 2-4: 327-347.

MAC ARTHUR, R.H. & E.O. WILSON (1967): The theory of island biogeography. Princeton Univ. Press, Princeton.

MANN, G. (1968): Die Ökosysteme Südamerikas. In: Biogeography and Ecology in South America. I: 171-229.

MAPS OF HISPANIC AMERICA 1: 1 000 000 (1942-1944): Index I-XII. U.S. Gov. Print. Office, Washington, D.C..

MAULL, O., KÜHN, F., TROLL, K. & W. KNOCH (1930): Südamerika in Natur, Kultur und Wirtschaft. Handbuch der Geographischen Wissenschaften, Wildpark-Potsdam.

MAYR, E. (1964): What is a fauna? Zool. Jb. Syst. 92: 473-486 (1965).

MAYR, E. (1966): Animal species and evolution. The Belknap Press, Harvard University Press, Cambridge, Mass., 3. print.

MAYR, E. (1969a): Principles of systematic zoology. McGraw-Hill Book Company, New York.

MAYR, E. (1969b): Bird speciation in the tropics. Biol. J. Linn. Soc. 1: 1-17.

MAYR, E., LINSLEY E.G. & R.L. USINGER (1953): Methods and principles of systematic zoology. McGraw-Hill Book Company, New York.

MAYR, E. & W. PHELPS (1967): The origin of the bird fauna of the South Venezuelan Highlands. Bull. Amer. Mus. Nat. Hist. 136: 273-327

MEISE, W. (1969): Seglervögel und Kolibris. In: Grzimeks Tierleben, VIII, Vögel 2, Kindler Verlag, Zürich.

MELLO-LEITÃO, C. DE (1946): As zonas de fauna de América tropical. Rev. Bras. Geograf. I.

MELVILLE, H. (1946): Die verfluchten Inseln. Basel.

MEYERS KONTINENTE UND MEERE (1969): Mittel- und Südamerika. Bibliographisches Inst., Mannheim.

MONROE, B.L.Jr. (1968): A distributional survey of the birds of Honduras. Americ. Ornith. Union, Allen Press, Lawrence.

MOORE, S. (1955): An annotated list of the moths of Michigan exclusive of Tineoidea (Lepidoptera). Misc. Publ. Mus. Zool. Univ. Michigan, 88.

MOOSER, O. (1940): Fauna Mexicana. III. (Insecta Lepidoptera, familia Sphingidae). Enumeración de los esfingidos Mexicanos (Insecta Lepidoptera), con notas sobre su morfología y su distribución en la República. Anales Escuela Nacional Ciencias Biológicas (Mexico) I: 407-494.

MOOSER, O. (1942): Esfingidos nuevos de Mexico (Lepidoptera, Sphingidae). Anales del Instituto de Biologia, 13: 205-211.

MORGAN, H. (1920): Dermaptera and Orthoptera. Calif. Acad. Sci. Proc. 2 (2): 311-346.

MOSS, A.M. (1912): On the Sphingidae of Peru. Trans. Zool. Soc. London, 20: 73-134.

MOSS, A.M. (1920): Sphingidae of Para, Brazil. Novit. Zoologicae, 27: 333-424.

MÜLLER P. (1970): Die Ausbreitungszentren terrestrischer Vertebraten in der Neotropis. Habilitationsschrift. Saarbrücken.

MÜLLER, P. (1971): Ausbreitungszentren und Evolution in der Neotropis. Mittl. Biogeogr. Abt. Univ. Saarl. 1: 1-20.

MÜLLER, P. (1972): Biogéographie et évolution en Amérique du Sud. Int. Geogr. Congr. Montreal.

MÜLLER, P. (1973): The dispersal centres of terrestrial vertebrates in the Neotropical realm. (Habilitationsschrift 1970, english edition). Biogeographica 2. Junk, The Hague.

MÜLLER, P. & J. SCHMITHÜSEN (1970): Probleme der Genese südamerikanischer Biota. In: Deutsche Geographische Forschung in der Welt von Heute. Festschr. Gentz: 109-122. Hirt Verlag, Kiel.

MÜLLER, P. & H. SCHREIBER (1972): Erfassung der Europäischen Wirbellosen. Mittl. Biogeogr. Abt. Univ. Saarl. 2: 1-12.

MURRAY, A. (1866): The geographical distribution of mammals. London.

NEEF, E. (1968): Das Gesicht der Erde. Taschenbuch der Physischen Geographie. Harri Deutsch, Frankfurt/M.-Zürich.

NETOLITZKY, F. (1939): Abfassung, Sammlung und Katalogisierung der Verbreitungskarten zur Insektengeographie. Verh. VII. Int. Kongr. Ent. Berlin (1938) I: 329-331.

NICHOLS, D. (ed.) (1962): Taxonomy and geography. Syst. Assoc. Publ. 4.

OITICICA, F.J. (1939): Sphingidae. In: Relatório excursão cientifica Instituto Oswaldo Cruz. Bol. Biol. (N.S.) 4 (2): 269-277.

OITICICA, F.J. (1942): Sphingidae capturados em Porto Cabral (Margem Paulista do Rio Parana), com notas sobre nomenclatura. Papéis Avulsos II (5): 97-102.

OITICICA, F.J. (1946): Revisão dos nomes genéricos da familia Sphingidae (Lepidoptera). Bol. Mus. Nac. (n.s.) Rio de Janeiro, Zoologia 66: 1-57.

ORLOG, C.C. (1969): Birds of South America. In: Biogeography and Ecology in South America. II: 849-878.

ORFILA, R.N. (1933): Estudios de Lepidopterologia Argentina III.-Catálogo sistematico de los Sphingidae (Lep.). Rev. Soc. Ent. Argent. 23: 189-206.

RAMBO, B. (1958): A historical approach to plant evolution. Pesquisas. Annuário do Instituto Anchietanto de Pesquisas 2: 257-298.

REBEL, H. (1931): Zur Frage der europäischen Faunenelemente. Ann. Nat. Hist. Mus. Wien 46.

REINIG, W.F. (1937): Die Holarktis. Ein Beitrag zur diluvialen und alluvialen Geschichte der zirkumpolaren Faunen- und Florengebiete. Fischer Verlag, Jena.

REINIG, W.F. (1938): Elimination und Selektion. Eine Untersuchung über Merkmalsprogressionen bei Tieren und Pflanzen auf genetisch- und historisch-chorologischer Grundlage. Fischer Verlag, Jena.

REINIG, W.F. (1950): Chorologische Voraussetzungen für die Analyse von Formenkreisen. In: Syllegomena biologica. Festschr. Kleinschmidt: 346-378.

ROESLER, U. (1965): Chorologische Untersuchungen über den *Homoeosoma-Ephestia*-Komplex (Lepidoptera: Phycitinae) im paläarktischen Raum. Bonn. Zool. Beitr. 16.

ROTHSCHILD, W. LORD & K. JORDAN (1903): A revision of the lepidopterous family Sphingidae. Novit. Zoologicae, IX, Suppl..

ROTHSCHILD, W. LORD & K. JORDAN (1910): List of the Sphingidae collected by the late W. Hoffmanns at Allianca, Rio Madeira, Amazonas. Novit. Zoologicae, 17 (3): 447-455.

ROTHSCHILD, W. LORD & K. JORDAN (1916): Further corrections of and additions to our revision of the Sphingidae. Novit. Zoologicae, 23 (2): 247-263.

RYAN, R.M. (1963): The biotic provinces of Central America. Acta Zool. Mex. 6 (2/3): 1-55.

SAVAGE, J.M. (1966): The origins and history of the Central American Herpetofauna. Copeia (4): 719-766.

SCHILDER, F.A. (1956): Lehrbuch der Allgemeinen Zoogeographie. Fischer Verlag, Jena.

SCHMIDT, K.P. (1954): Faunal realms, regions and provinces. Quart. Rev. Biol. 29: 322-331.

SCHMITHÜSEN, J. (1968): Allgemeine Vegetationsgeographie. Walter de Gruyter & Co., Berlin.

SCHMITHÜSEN, J. (1969): Vegetationskarte. In: Meyers Kontinente und Meere. Mittel- und Südamerika: 30-31.

SCHREITER, R. (1926): Sphingidae. Estudio sobre las especies tucumanas de esta familia. Univ. Nac. Tucuman 9.

SCLATER, P.L. (1858): On the general geographical distribution of the members of the class Aves. Journ. Proc. Linn. Soc. London (Zool.) 2: 130-145.

SEITZ, A. (1927): Sphingidae. Allgemeines. In: A. SEITZ: Die Gross-Schmetterlinge der Erde, 14: 353-357. Alfred Kernen Verlag, Stuttgart.

SEITZ, A. (1931): Sphingidae. Allgemeines. In: A. SEITZ: Die Gross-Schmetterlinge der Erde, 6: 839-845. Alfred Kernen Verlag, Stuttgart.

SELANDER, R.B. & P. VAURIE (1962): A gazetteer to accompany the 'Insecta' Volumes of the 'Biologica Centrali-Americana'. Am. Mus. Nov. 2099.

SIMPSON, G.G. (1953): Evolution and geography. Eugene, Oregon.

SIMPSON, G.G. (1964): Species density of North American mammals. Syst. Zool. 13: 57-73.

SIMPSON, G.G. (1965): The geography of evolution. Chilton Company, Book Division. Philadelphia.

SIOLI, H. (1968): Zur Ökologie des Amazonasgebietes. In: Biogeography and Ecology in South America 1: 137-170.

SMITH, H.M. (1940): An analysis of the biotic provinces of Mexico, as indicated by the distribution of the lizards of the genus *Sceloporus*. Anales Esc. Nac. Cienc. Biol. 2 (1): 55-110.

THE TIMES ATLAS OF THE WORLD (1967) Houghton Mifflin Company, Boston.

TIETZ, H.M. (1972): An index to the described life histories, early stages and hosts of the Macrolepidoptera of the Continental United States and Canada. Allyn Mus. Ent. Sarasota.

TORRE, S.L. DE LA (1960): Estudio de los órganos genitales de las Sphingidae de Cuba contenidas en la colección de la Universidad de Oriente. Rev. Univ. Oriente I: 41-75.

TROLL, C. (1955): Der jahreszeitliche Ablauf des Naturgeschehens in den verschiedenen Klimagürteln der Erde. Stud. Gen. 8.

UDVARDY, M.D.F. (1969): Dynamic Zoogeography. Van Nostrand Reinhold Company. New York.

UNVARDY, M.D.F. (1975): A classification of the biogeographical provinces of the world. IUCN Occasional Paper 18, Morges.

URETA, E.R. & R.B. DONOSO (1956): Revisión de la familia Sphingidae (Lep. Het.) en Chile. Bol. Mus. Nac. Hist. Nat. Chile (26): 237-256.

URQUART, F.A. (1960): The monarch butterfly. University of Toronto Press, Toronto.

VAVILOV, N.J. (1926): Studies on the origin of cultivated plants. Bull. Appl. Bot. Plant. Breed., Leningrad.

VAVILOV, N.J. (1931): Origin of cultivated plants. Bull. Appl. Bot. Genet. Leningrad.

VOGL, P.C. (1944): Esfingidos (Sphingidae) y Dipteros (Diptera) de la Hacienda la Trinidad de Maracay, coleccionados por P. CORNELIUS VOGL. Bol. Soc. Venez. Cienc. Nat. IX: 321-322.

VUILLEUMIER, B.S. (1971): Pleistocene changes in the fauna and flora of South America. Science 173 (3999): 771-780.

VUILLEUMIER, F. (1965): Relationship and evolution within the Cracidae (Aves/Galliformes). Mus. Com. Zool. 134 (1): 1-27.

VUILLEUMIER, F. (1970): Insular biogeography in continental regions. I. The Northern Andes of South America. The American Naturalist 104 (938): 373-388.

WAGNER, H.O. (1961): Die Nagetiere einer Gebirgsabdachung in Südmexiko und ihre Beziehungen zur Umwelt. Zool. Jb. Syst. 89: 177-242.

WALLACE, A.R. (1876): The geographical distribution of animals. 2 vols. Reprinted 1962, Hafner, New York and London.

WALTER, H. (1954): Grundlagen der Pflanzenverbreitung. 2. Arealkunde. Ulmer Verlag, Stuttgart.

WARNER, H.H. (1934): *Angraecum* and *Xanthopan*. The story of an orchid and a moth. In: Gardening Illustrated LVI (2887): 403.

WELLING, E.C. (1966): Devastation of Yucatecan forests, with notes on insect abundance and formation of local climates. J. Lepid. Soc. 20 (4): 201-205.

WEYL, R. (1956): Eiszeitliche Gletscherspuren in Costa Rica (Mittelamerika). Z. Gletscherkd. u. Glazialgeol. 3: 37-325.

WEYL, R. (1970): Mittelamerika. Krustenbau und paläographische Entwicklung. In: Umschau (10): 295-299.

WHEELER, W.M. (1919): The ants of the Galápagos Islands. The ants of Cocos Islands. Calif. Acad. Sci. Proc. 2: 259-308.

WILHELMY, H. (1952): Die eiszeitliche und nacheiszeitliche Verschiebung der Klima- und Vegetationszonen in Südamerika. Dt. Geographentag Frankfurt.

WILLIAMS, F.X. (1911): The butterflies and hawk-moths of the Galápagos Islands. Calif. Acad. Sci. Proc. 1: 289-322.

WILLIAMS, F.X. (1926): The bees and aculeate wasps of the Galápagos Islands. Calif. Acad. Sci. Proc. 2 (2): 347-357.

WOODRING, W.P. (1966): The Panama landbridge as a sea barrier. Proc. Am. Phil. Soc. 110 (6): 425-433.

ZAYAS M.F. & A.D. ALAYO (1956): La familia Sphingidae en Cuba (Lepidoptera: Heterocera). Univ. Oriente 40, Santiago de Cuba.

Author's address: HARALD SCHREIBER, Schwerpunkt Biogeographie, Universität des Saarlandes, 66 Saarbrücken.

IX. ANNEX

a. Range maps and locality guide-maps

The following range maps of Sphingidae which have been mentioned in the text have been selected from a total of 576 drawn range maps which will be published later.

The symbols used for mapping the localities differ in size according to the number of specimens recorded from one locality (1, 2-10, 11-30, >30). All localities from which sphingids were recorded have been entered and numbered in the accompanying locality guide-maps. With the aid of these numbers and the respective grid references as given in the locality index, each dot on the range maps marking a locality can be determined.

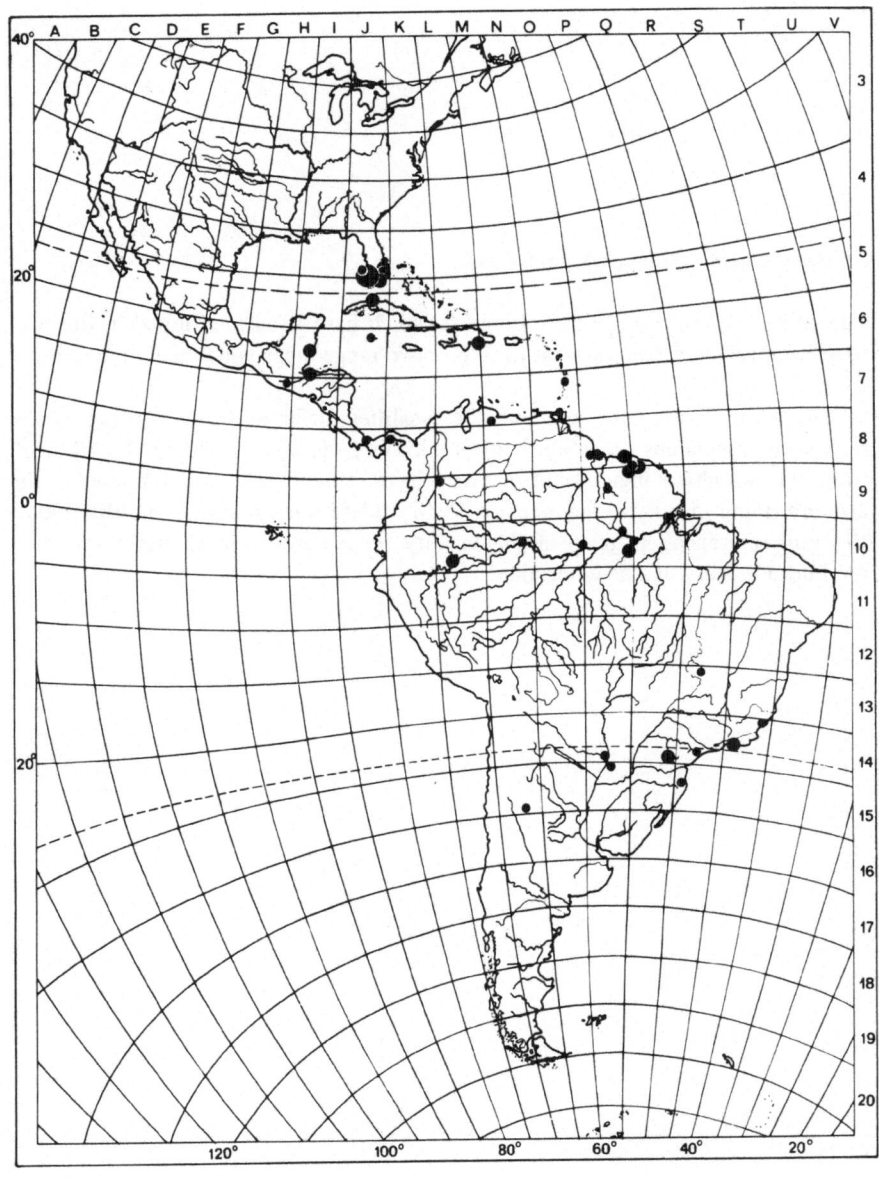

Fig. 30. Range of *Phryxus caicus* (CRAM.).

138

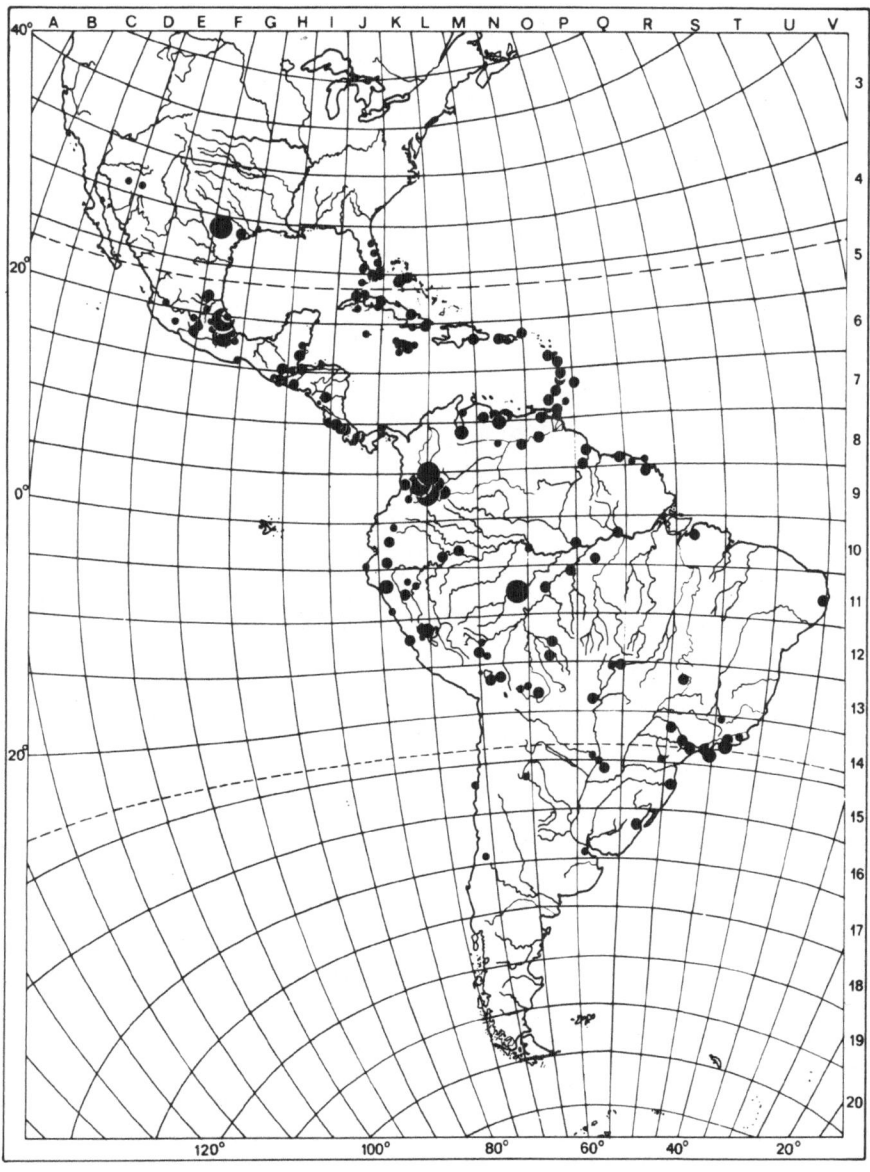

Fig. 31. Range of *Pachylia ficus* (L.).

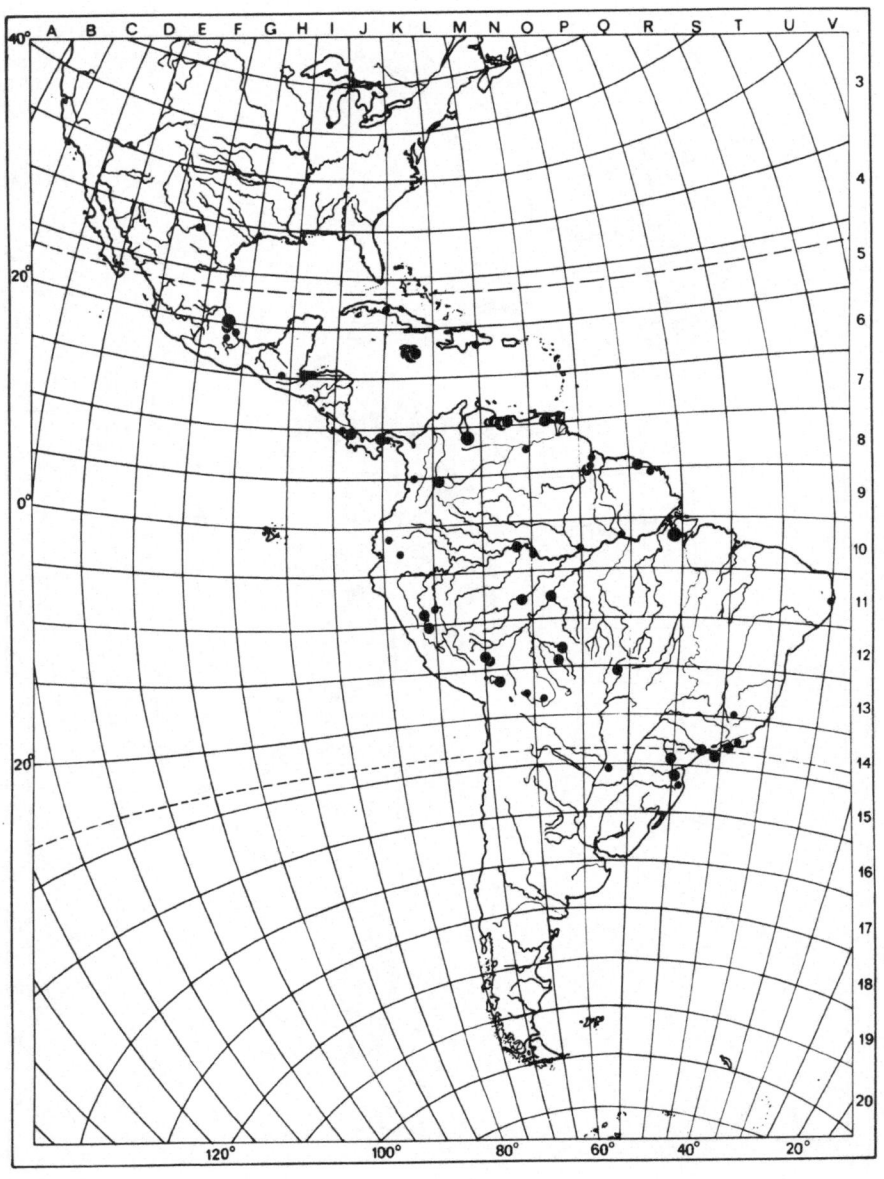

Fig. 32. Range of *Neococytius cluentius* (CRAM.).

140

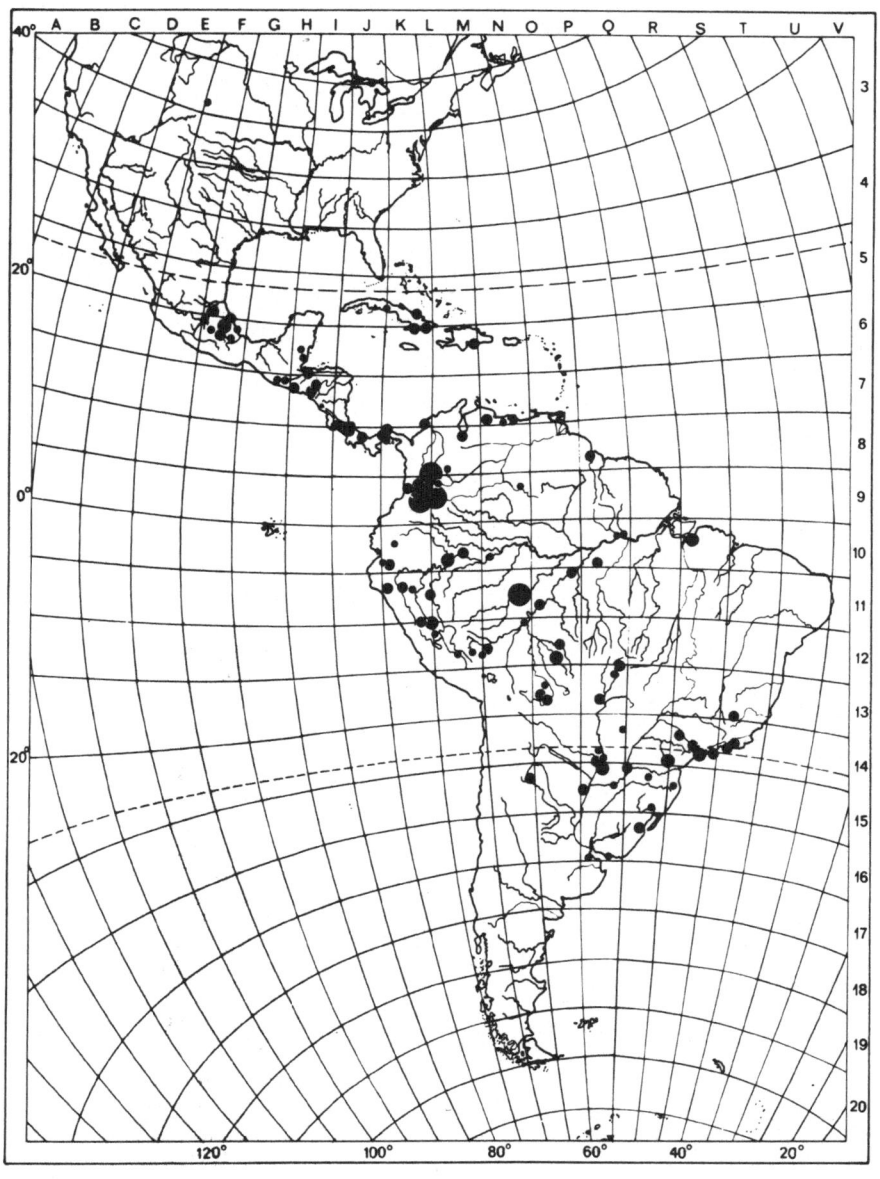

Fig. 33. Range of *Pachylioides resumens* (WLK.).

141

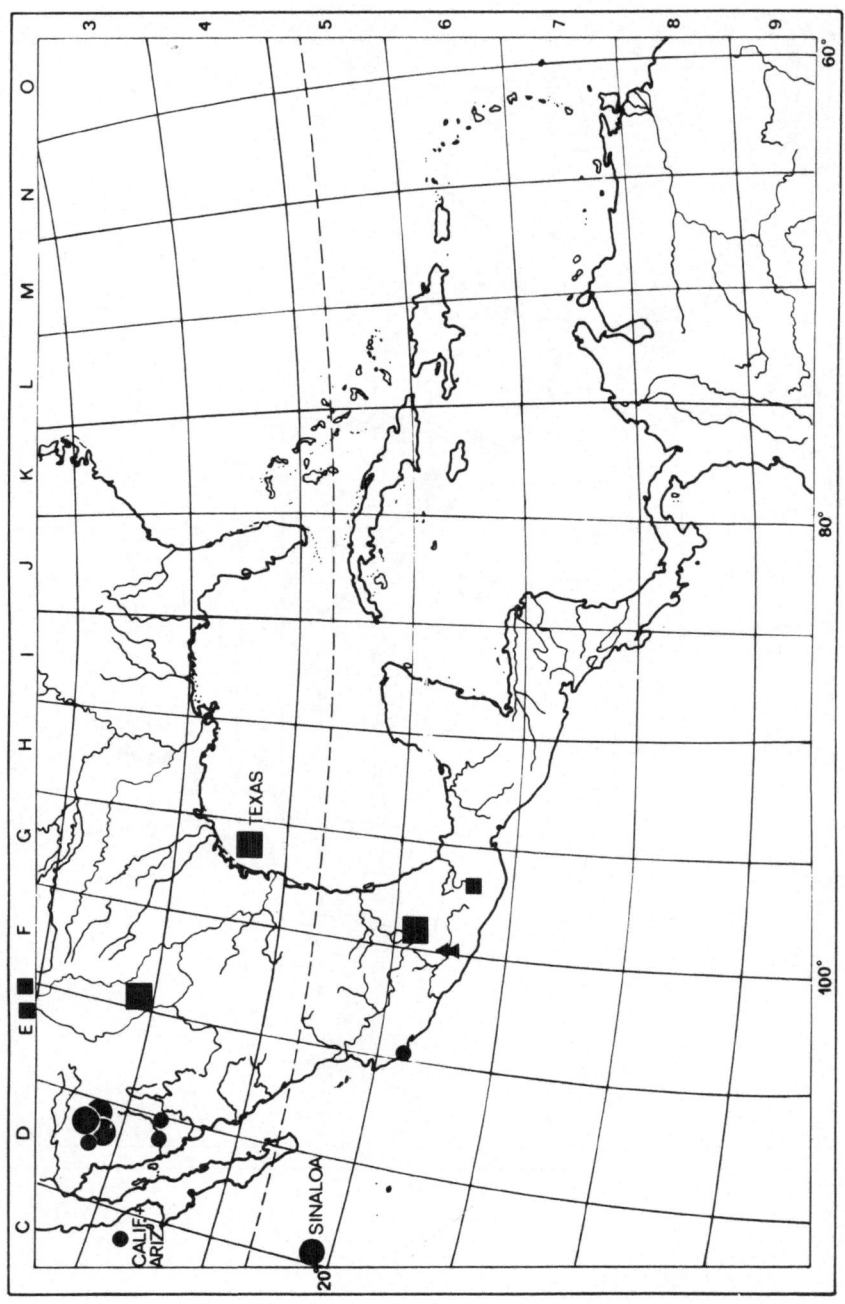

Fig. 34. Ranges of *Proserpinus vega* (DYAR) ■, *Proserpinus terlooii* HY. EDW ,
Proserpinus terlooii mooseri (CLARK) ▲.

142

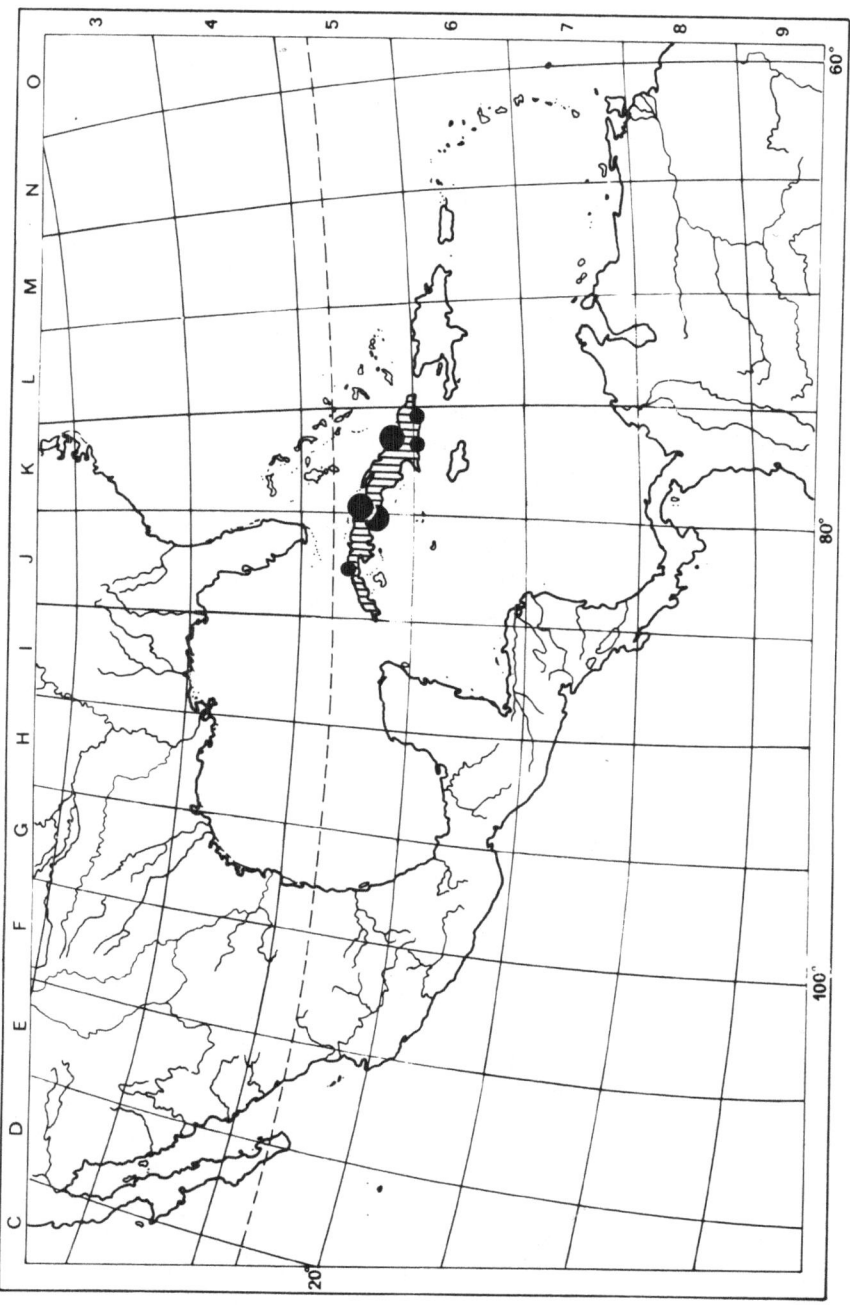

Fig. 35. Range of *Xylophanes gundlachi* (H.-S.).

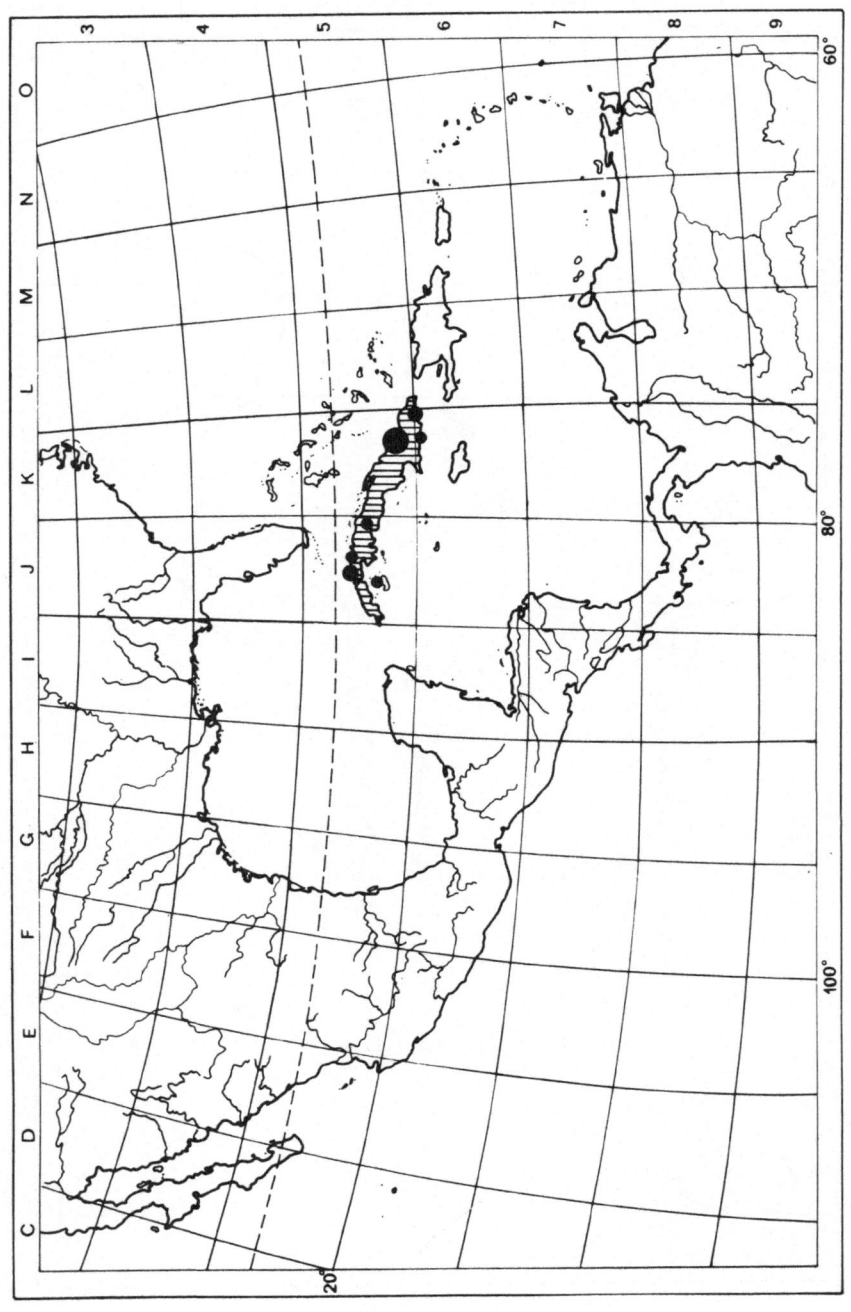

Fig. 36. Range of *Xylophanes robinsoni* (GRT.).

144

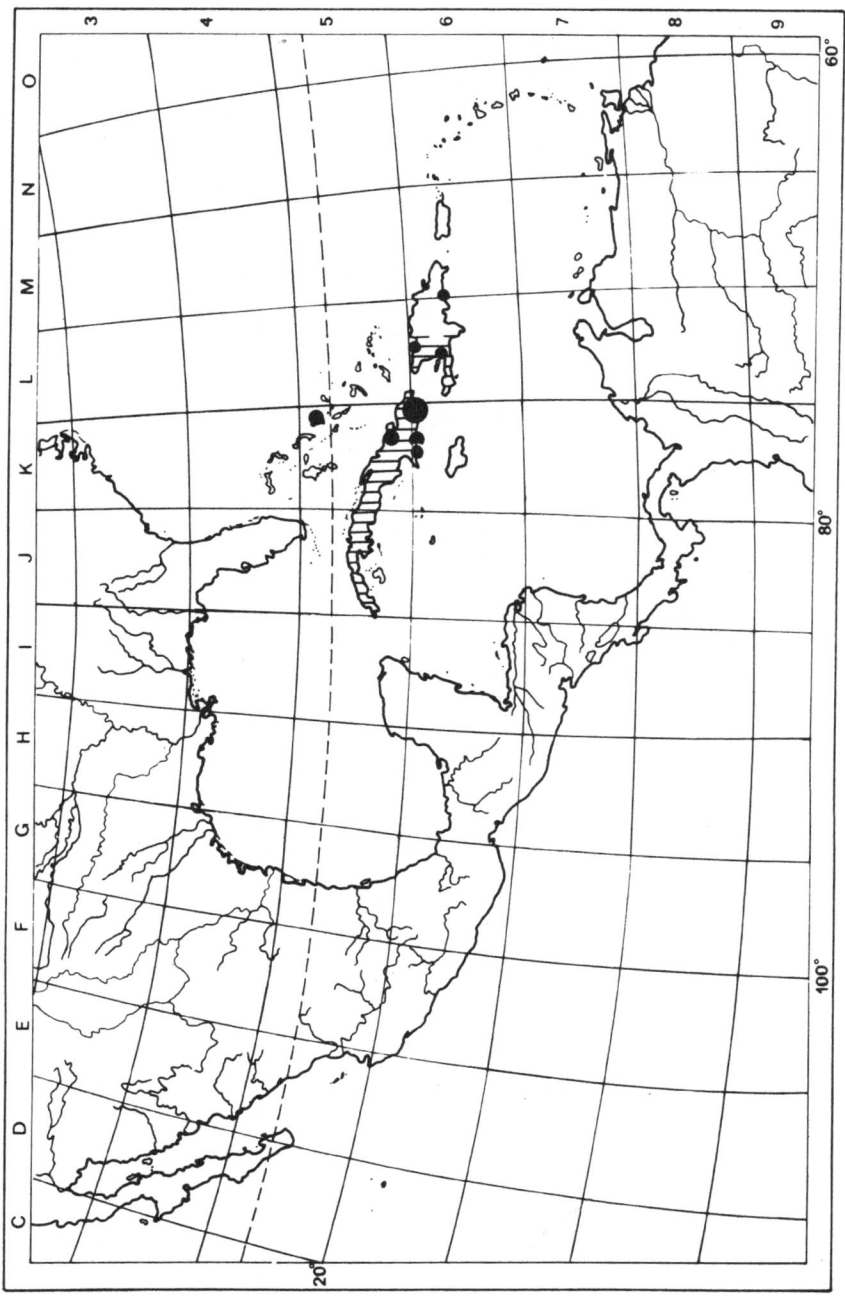

Fig. 37. Range of *Perigonia lefeburei* (LUC.).

145

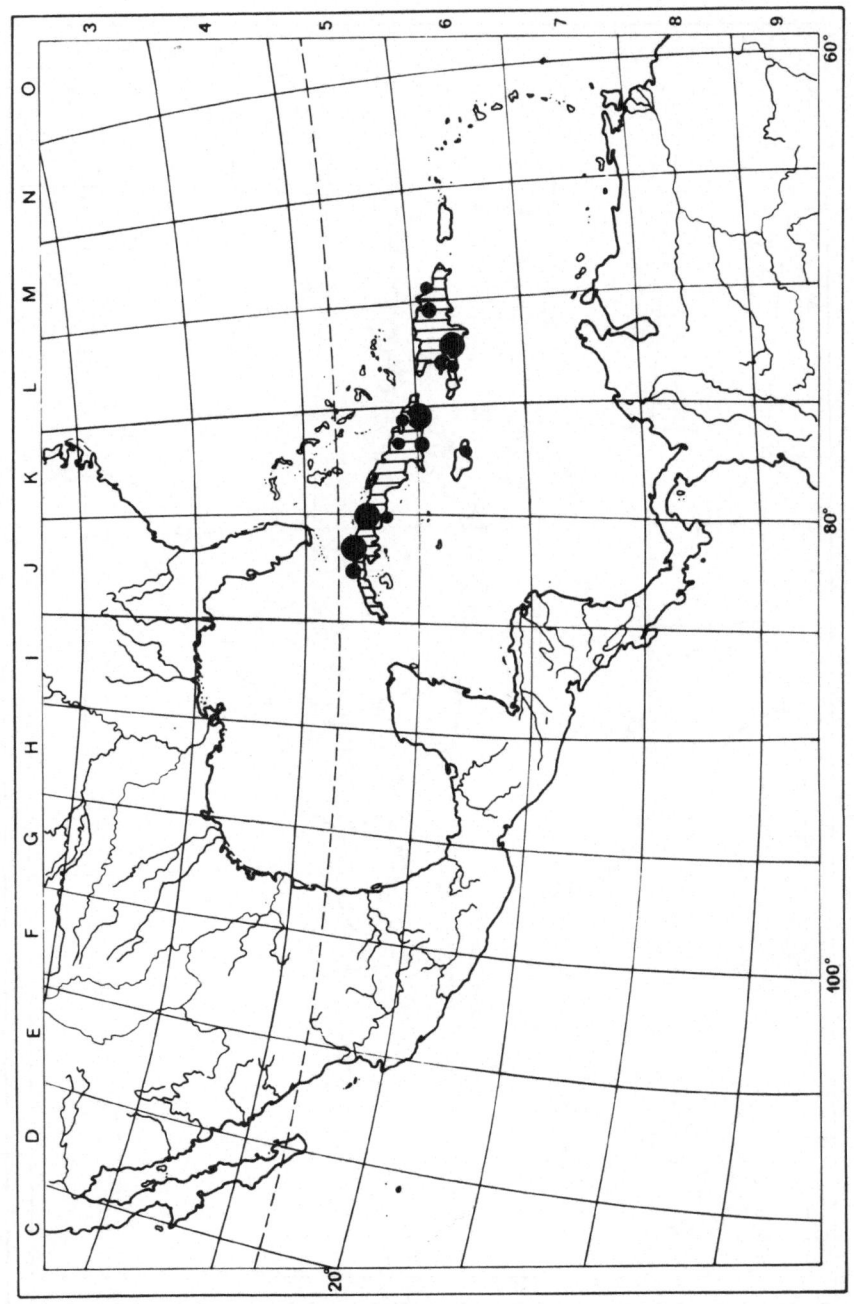

Fig. 38. Range of *Erinnyis guttularis* (WLK.).

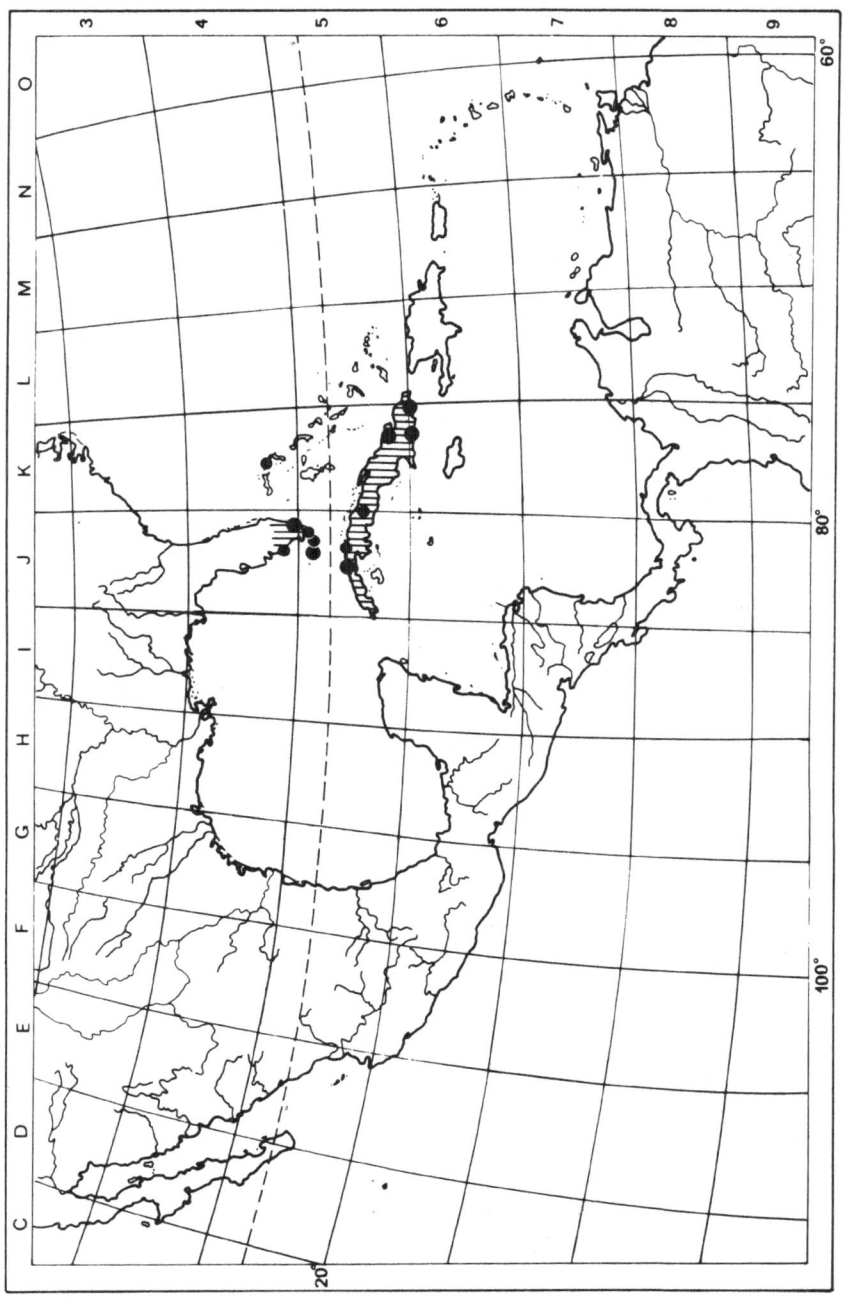

Fig. 39. Range of *Madoryx pseudothyreus* (GRT.).

147

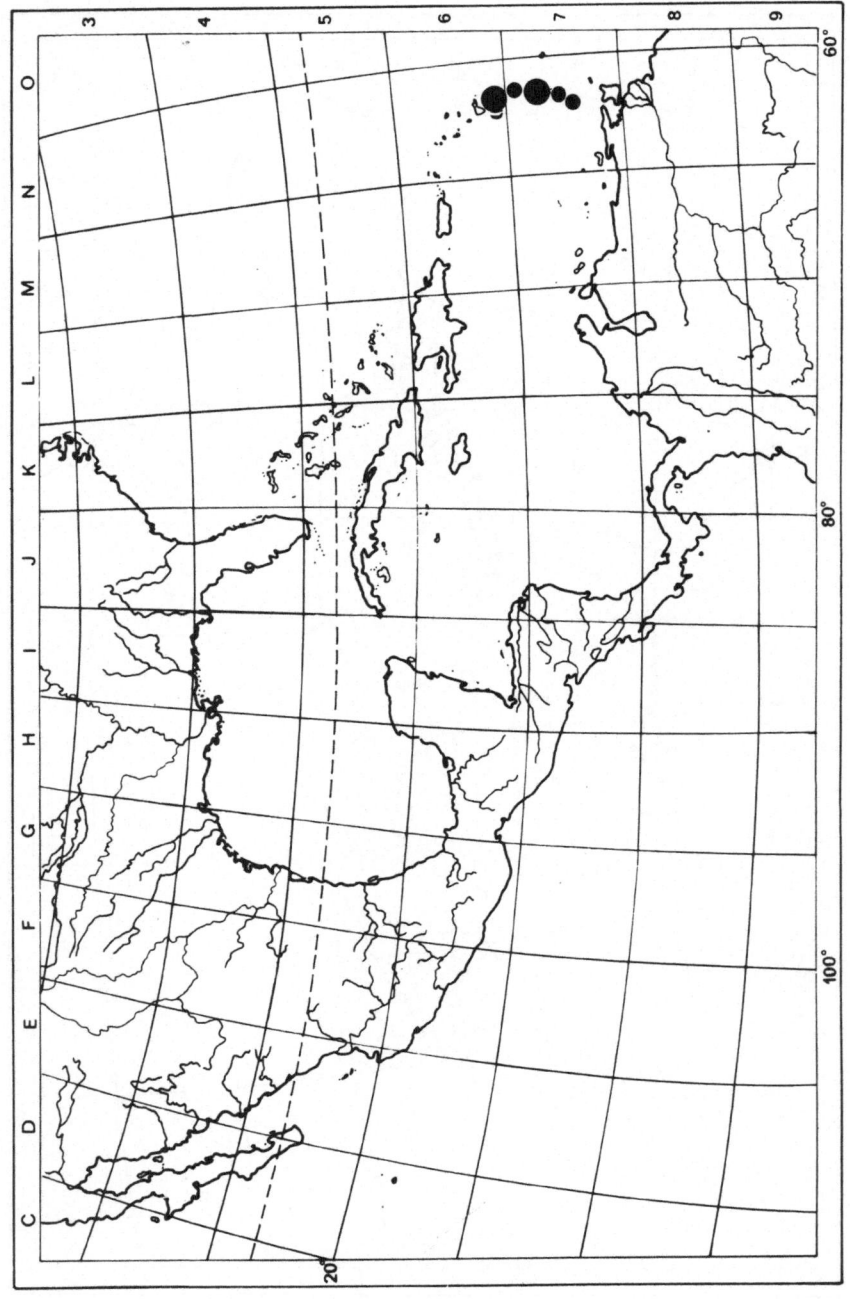

Fig. 40. Range of *Eumorpha vitis fuscata* (R. & J.).

148

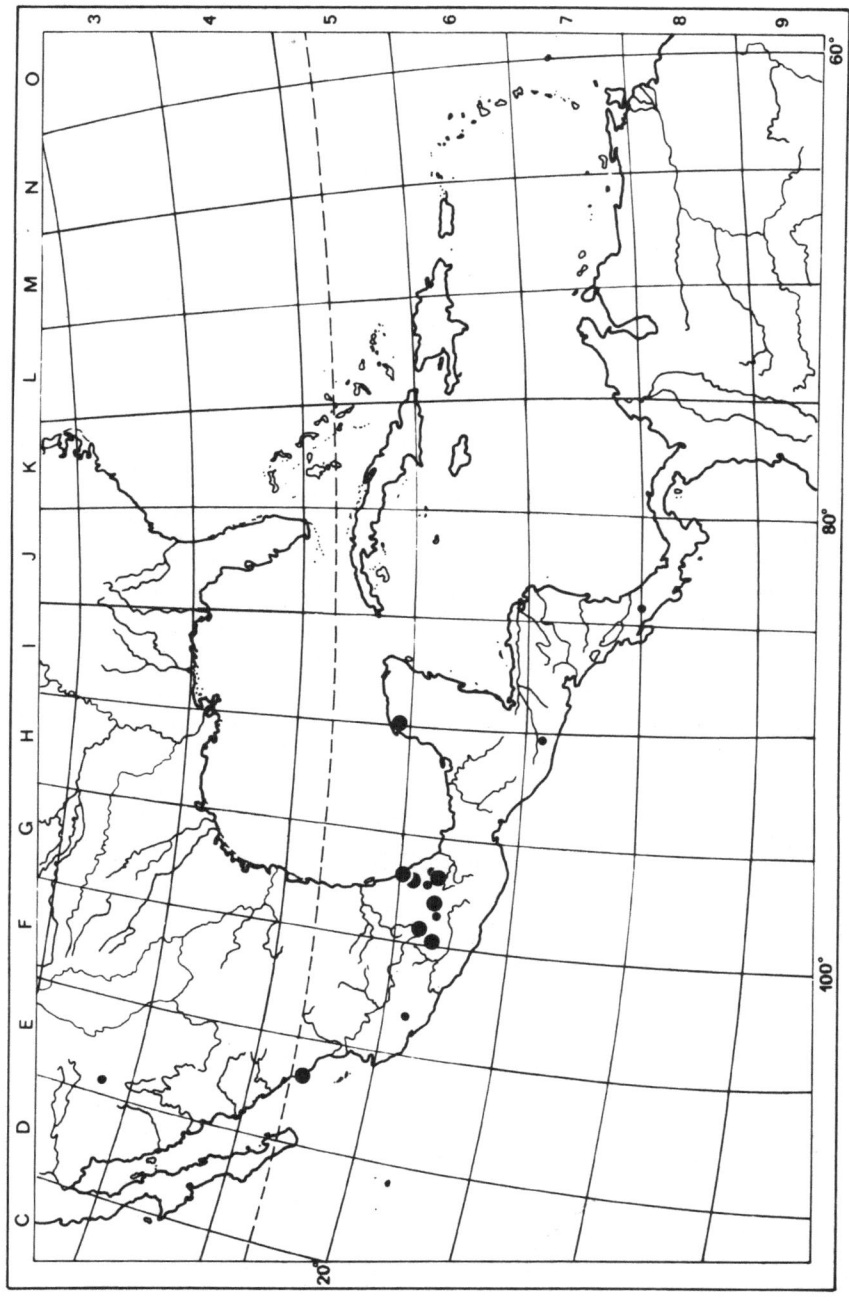

Fig. 41. Range of *Erinnyis yucatana* (DRC.).

149

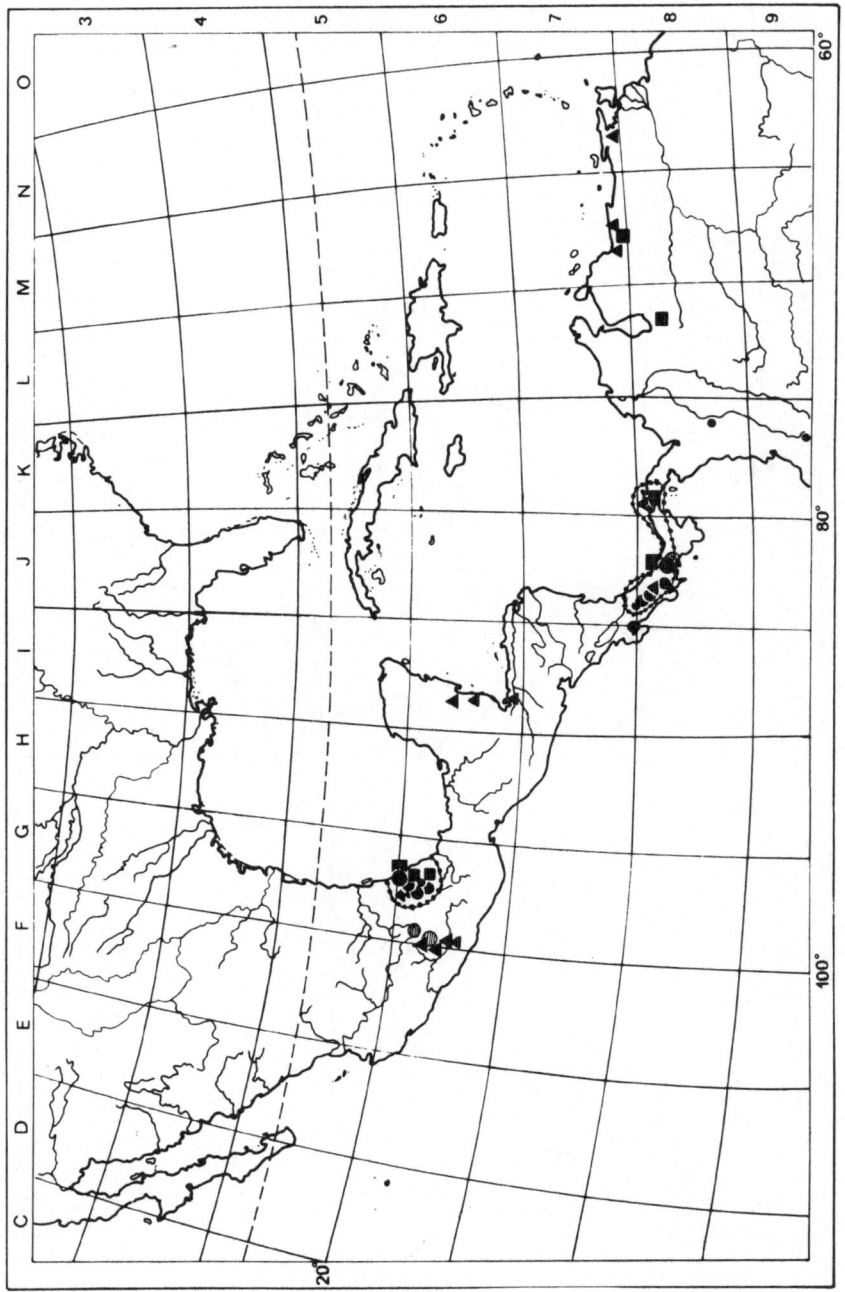

Fig. 42. Disjunct distribution between the centres in Mexico and the Costa Rican centre: *Amplypterus globifer* DYAR ⊜ , *Madoryx pluto dentatus* GEHLEN ●●●●, *Stolidoptera tachasara* (DRC.) ■, *Xylophanes turbata* (HY.EDW.) ▲, *Xylophanes amadis cyrene* (DRC.) ●.

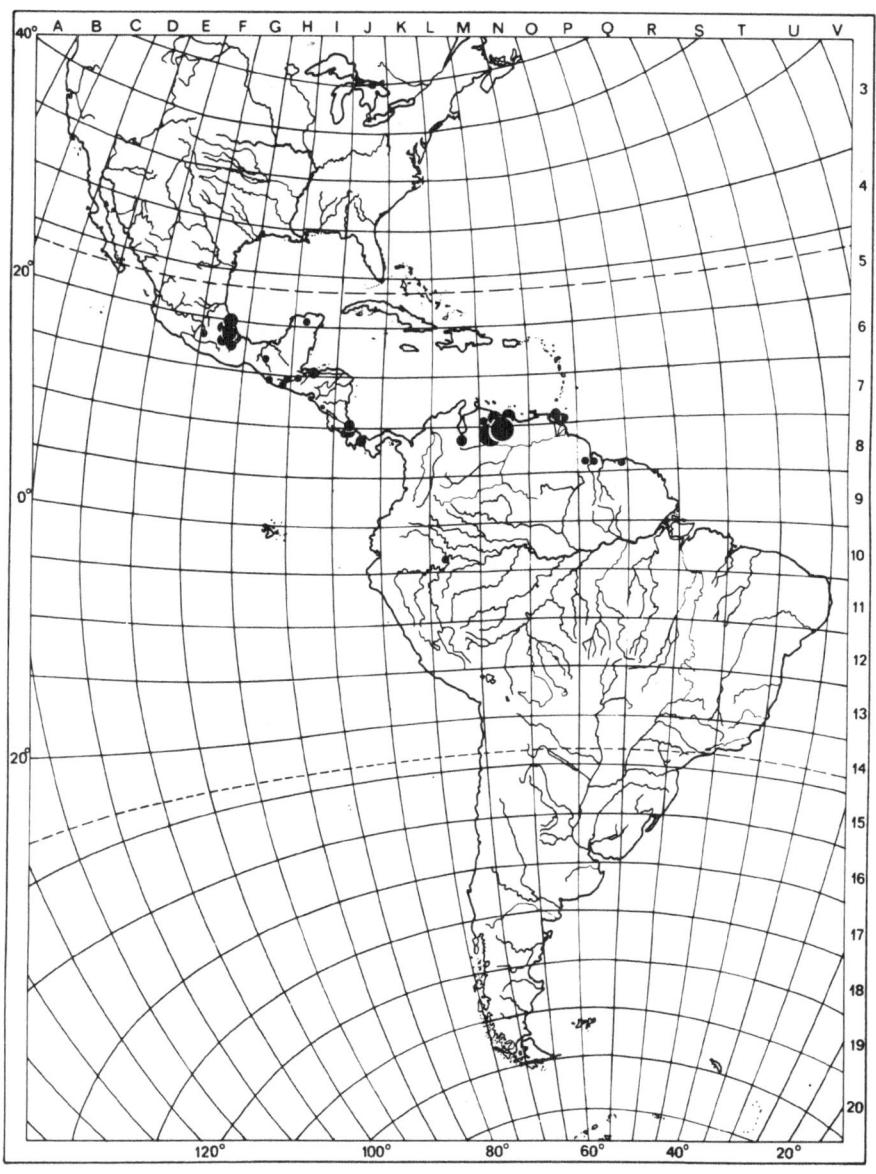

Fig. 43. Range of *Xylophanes neptolemus* (STOLL).

151

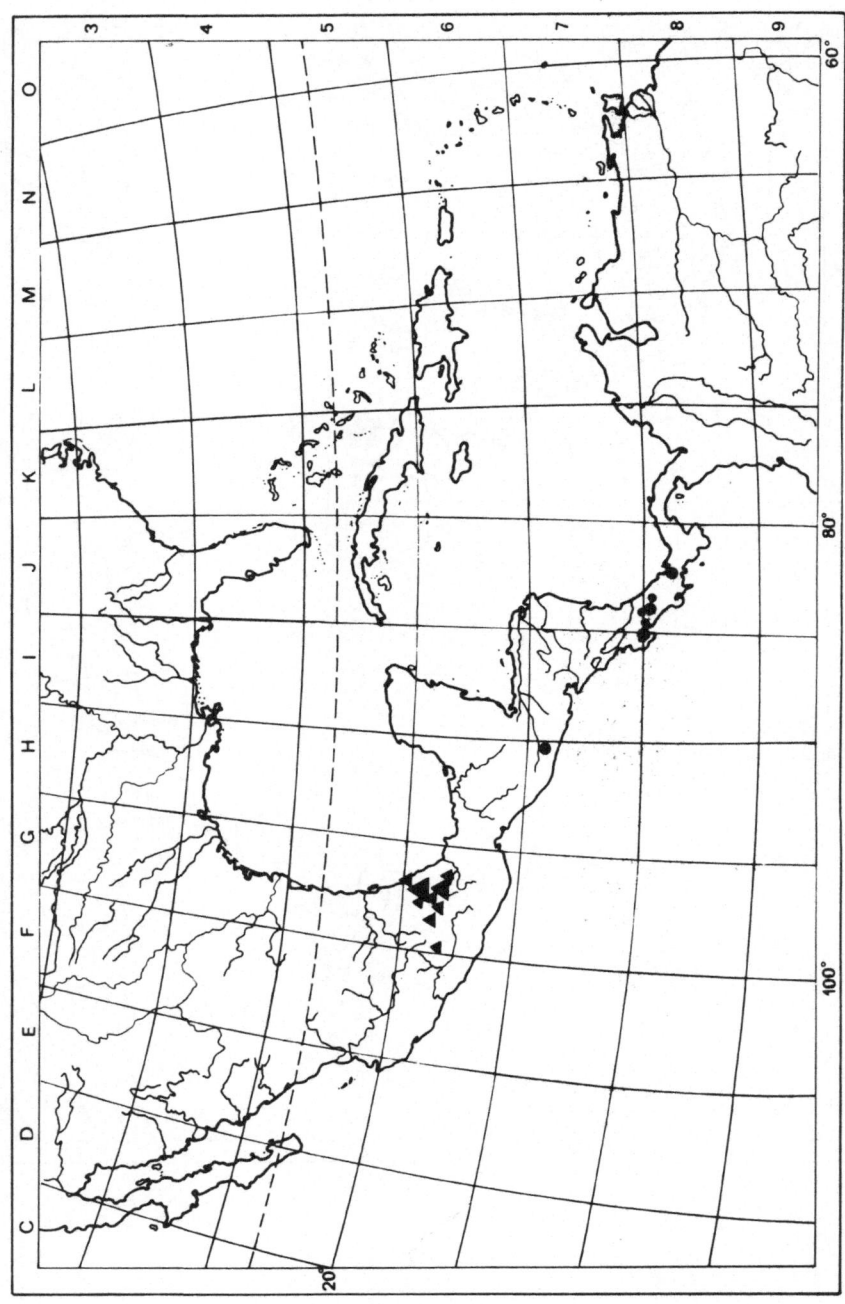

Fig. 44. Ranges of *Amplypterus donysa* (DRC.) ▲ and *Amplypterus donysa dariensis* R. & J. ●.

152

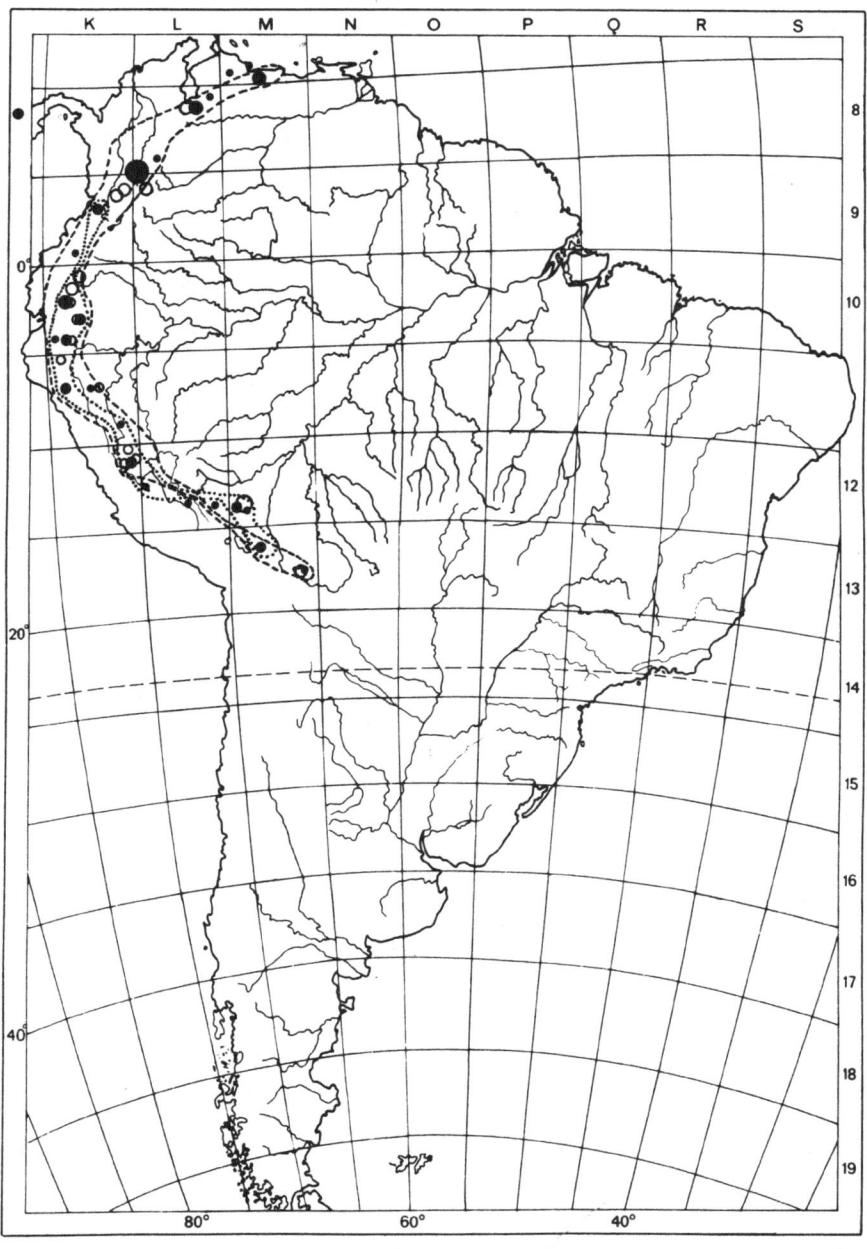

Fig. 45. Ranges of *Manduca scutata* (R. & J.) ●, *Manduca trimacula* (R. & J.) ....., *Euryglottis aper* (WLK.) ---, *Amplypterus sexoculata* (GRT.) ○.

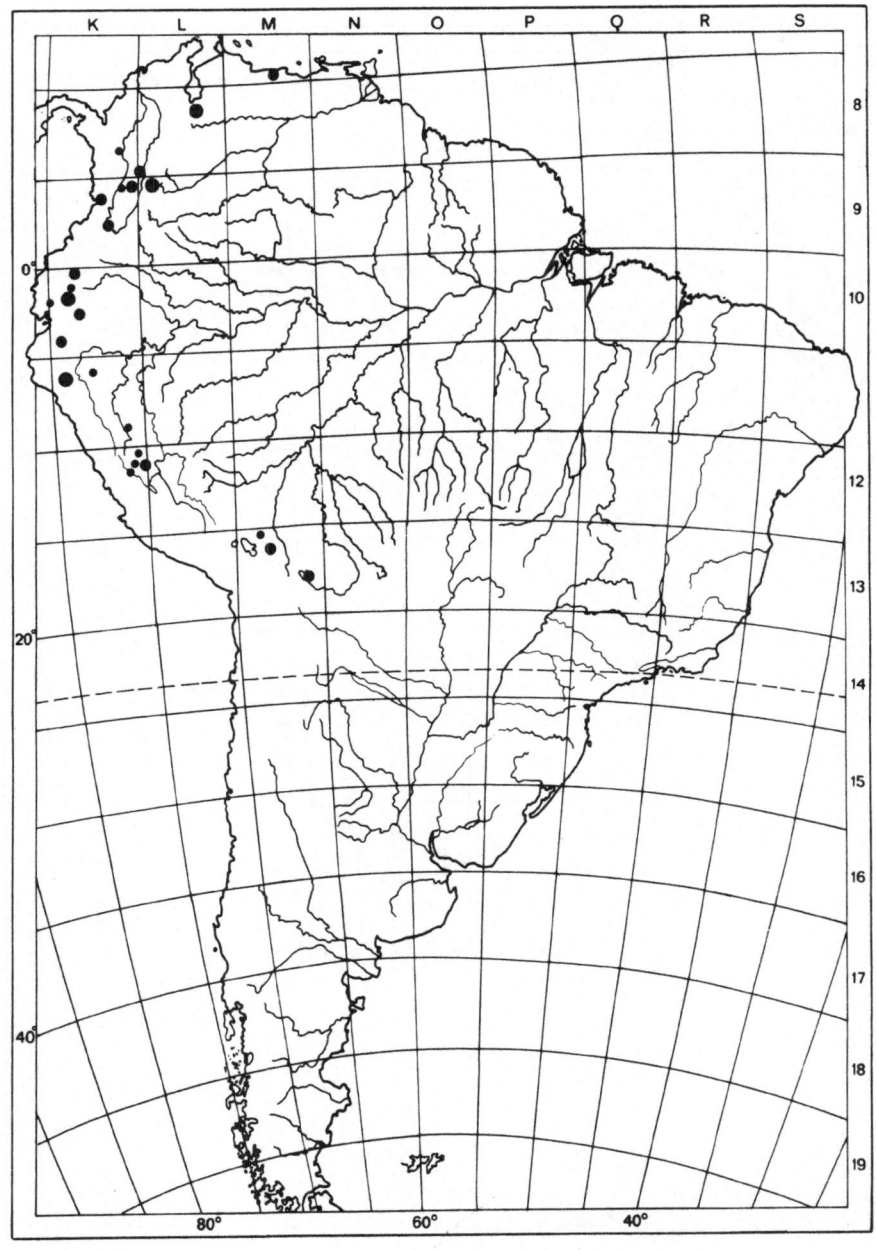

Fig. 46. Range of *Euryglottis aper* (WLK.).

154

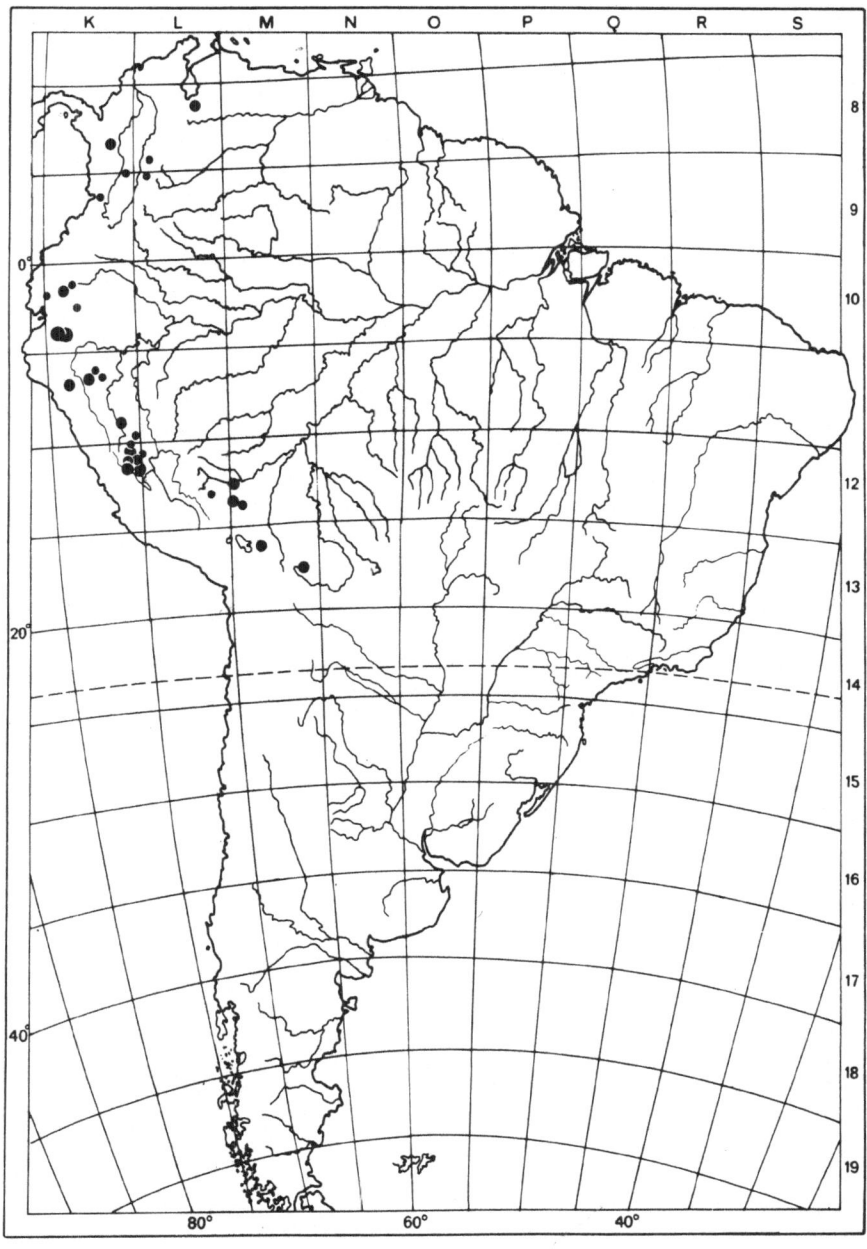

Fig. 47. Range of *Nyceryx hyposticta* (FLDR.).

155

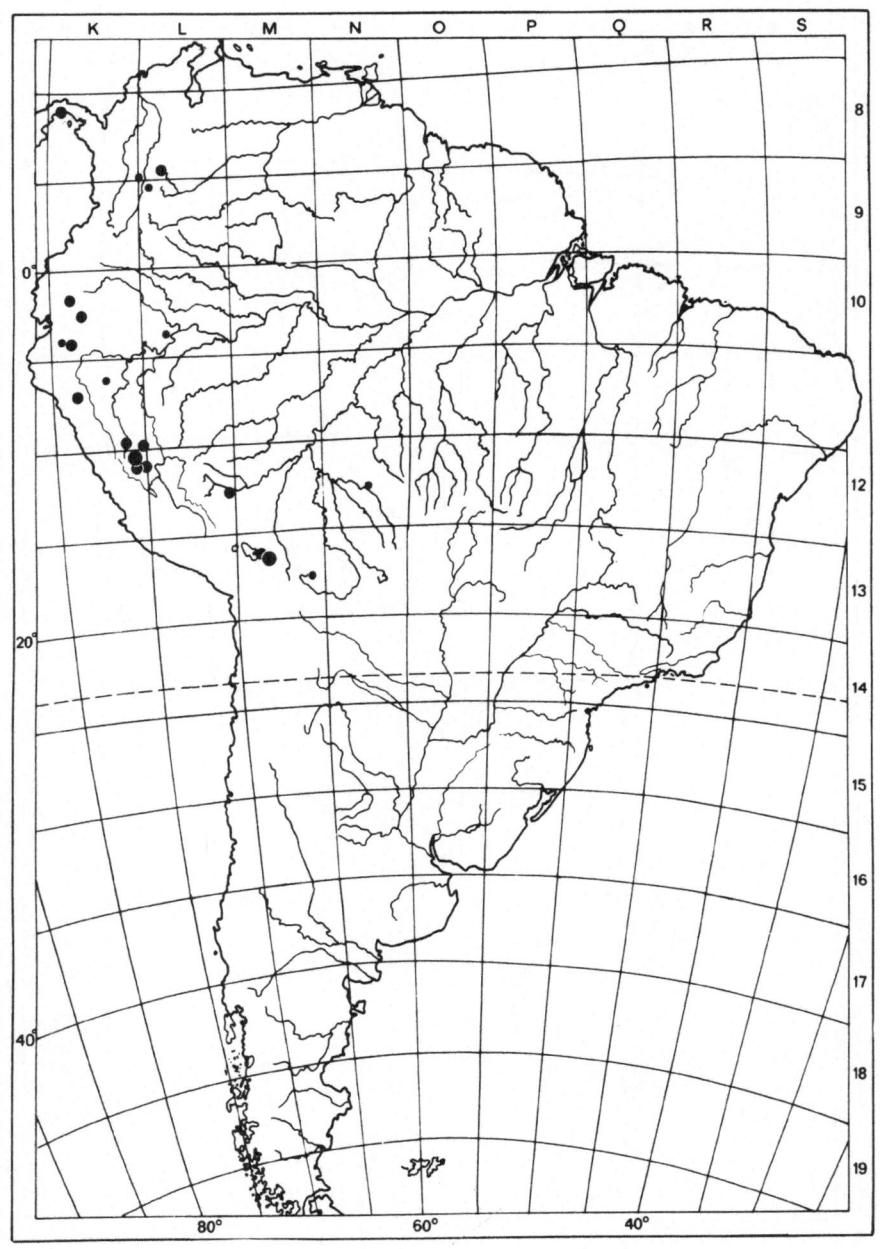

Fig. 48. Range of *Pachygonia hopfferi* STGR.

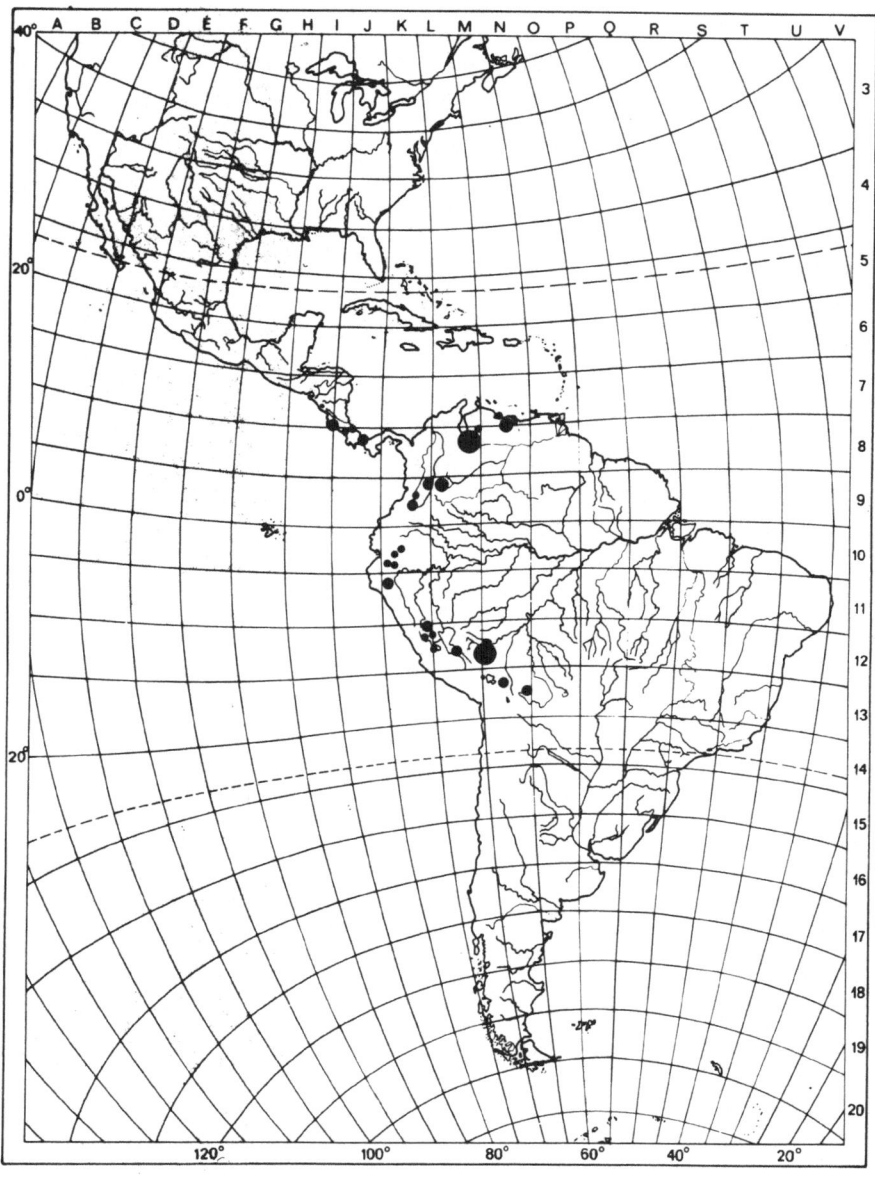

Fig. 49. Range of *Xylophanes crotonis* (WLK.).

157

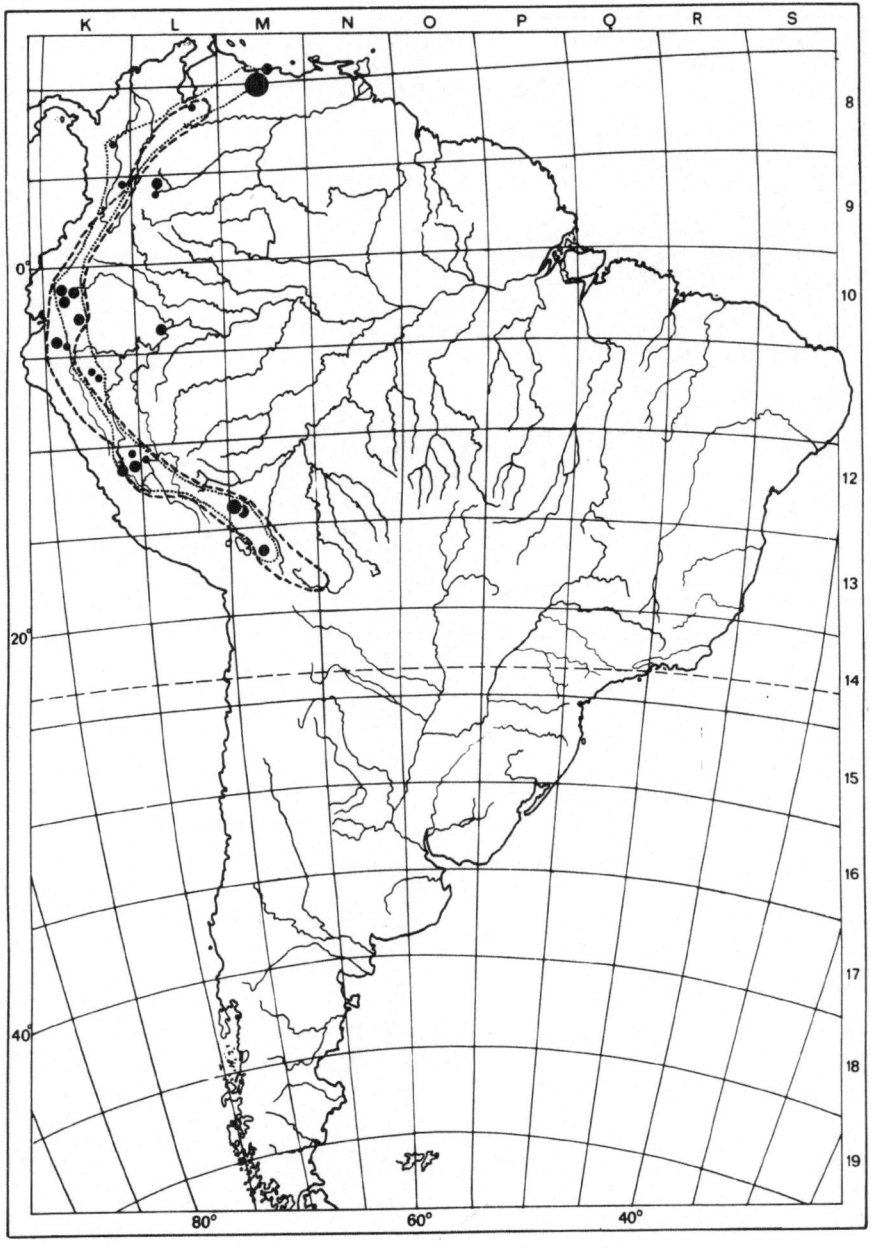

Fig. 50. Ranges of *Euryglottis dognini* ROTHSCH. —, *Protambulyx euryalus* R. & J. ....., *Amplypterus tigrina* (FLDR.) ●.

158

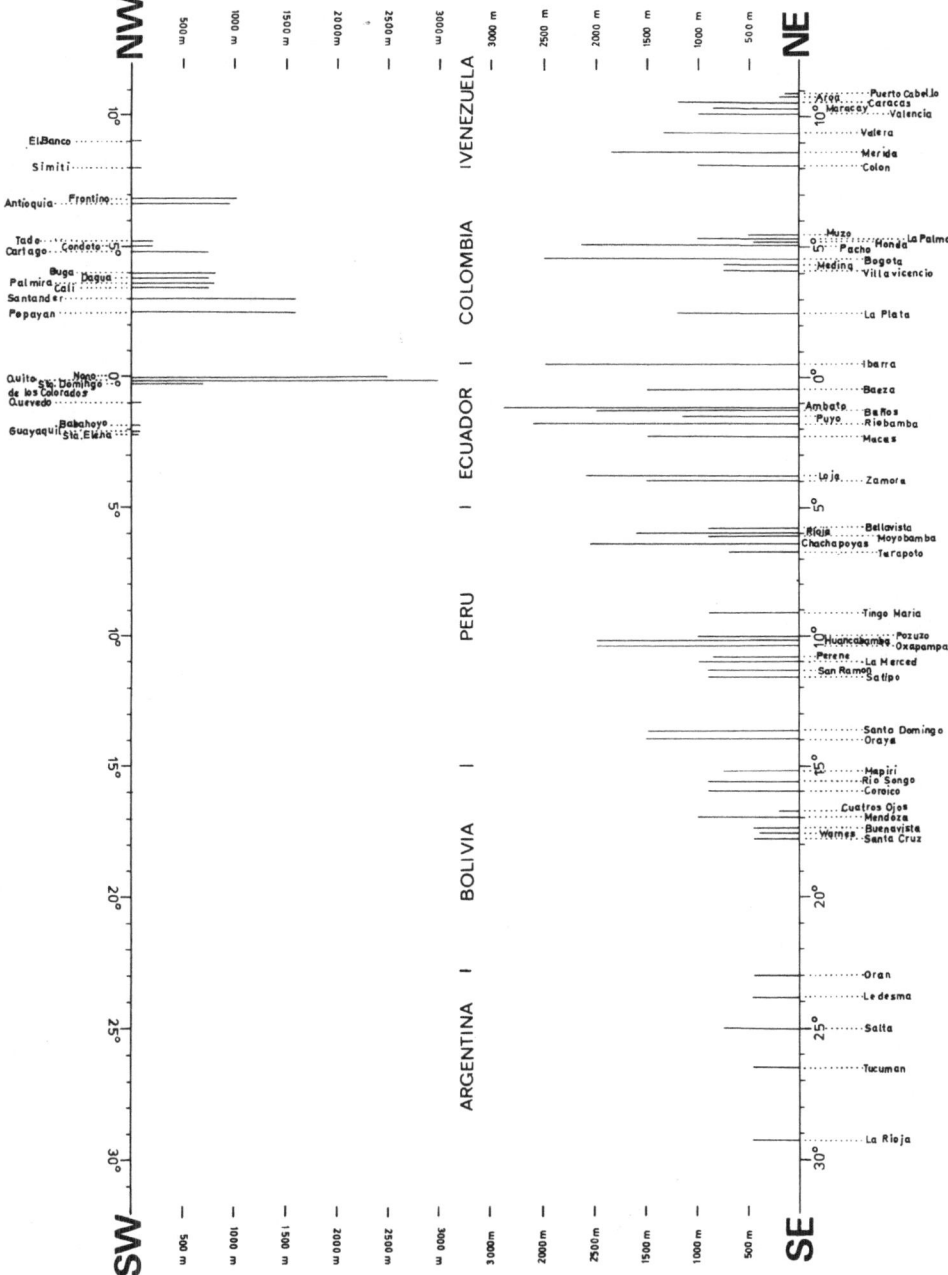

Fig. 51. Vertical distribution of sphingid localities in the Andes.

159

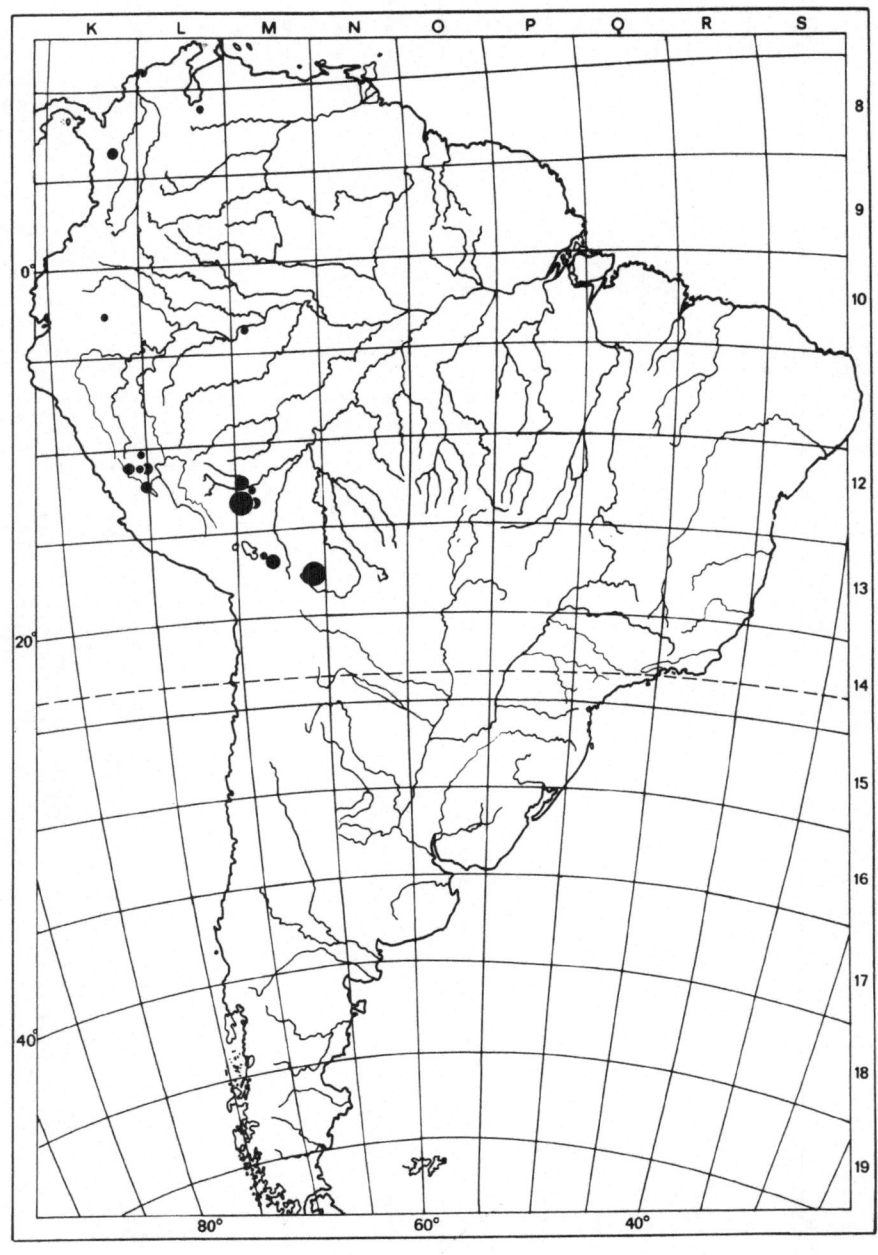

Fig. 52. Range of *Xylophanes pyrrhus* R. & J.

160

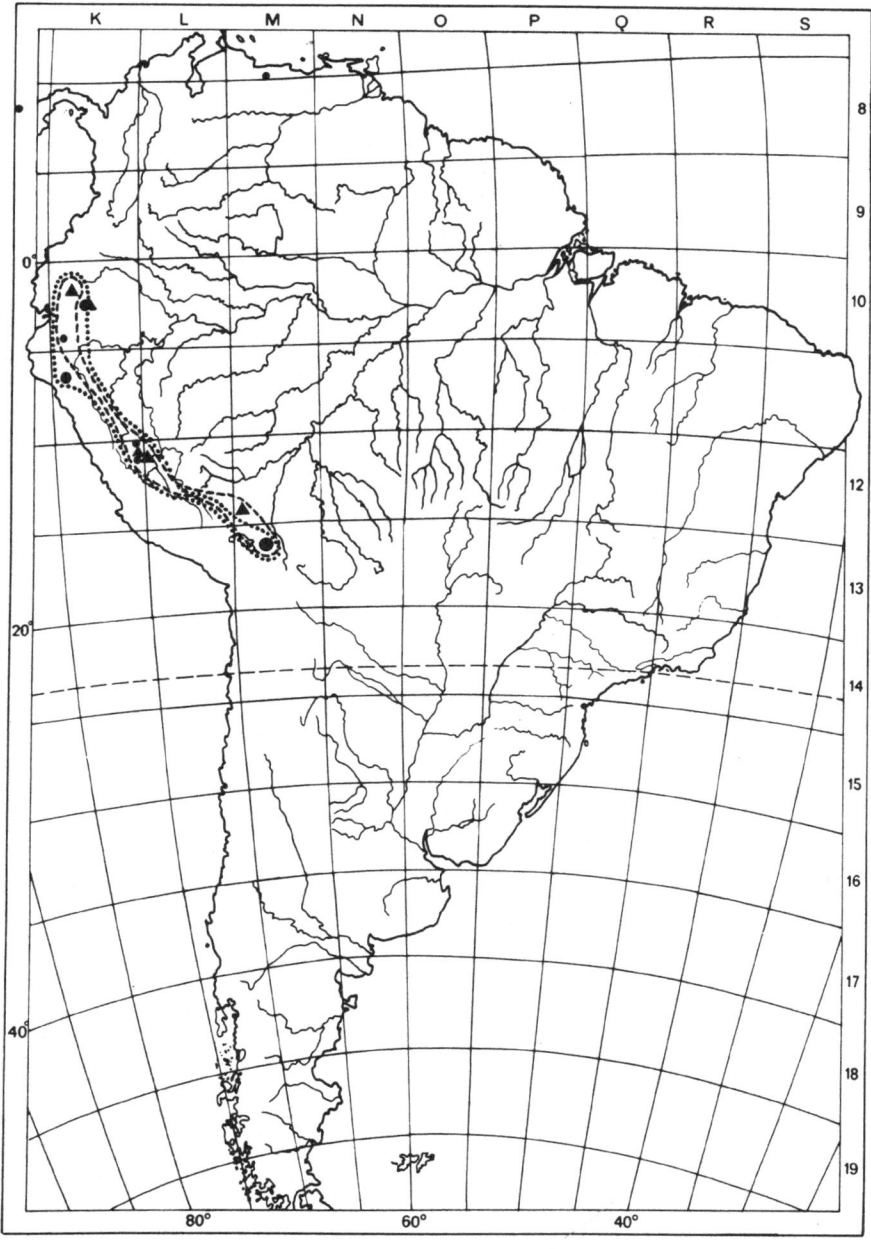

Fig. 53. Ranges of *Manduca andicola* (R. & J.) - - -, *Manduca extrema* (GEHLEN) ●, *Euryglottis albostigmata basalis* ROTSCH. ▲, *Nyceryx nictitans saturata* R. &. J. ⋯⋯ .

161

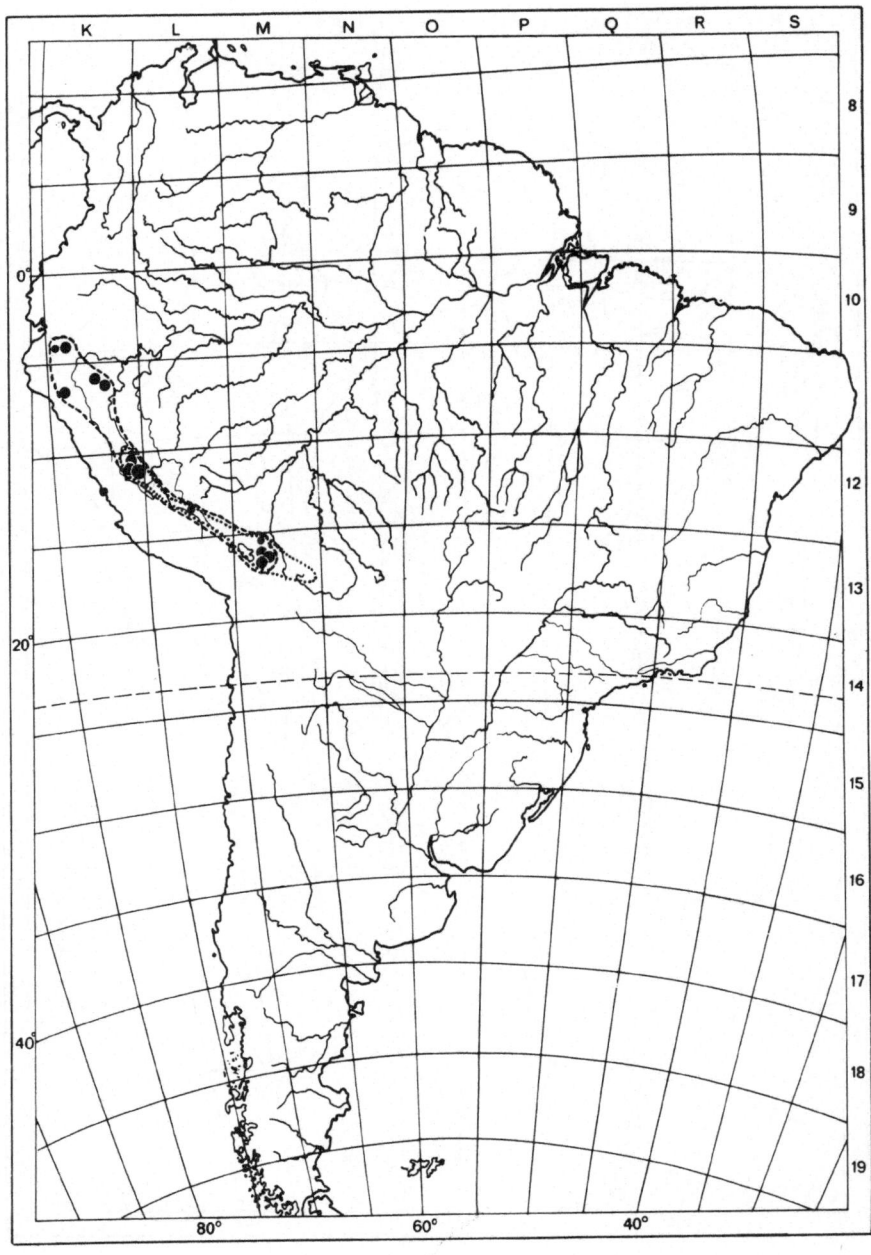

Fig. 54. Ranges of *Sphinx aurigutta* (R. & J.) ·····, *Perigonia grisea* R. & J. ●, *Eupyrrhoglossum corvus* (BDV.) --- .

162

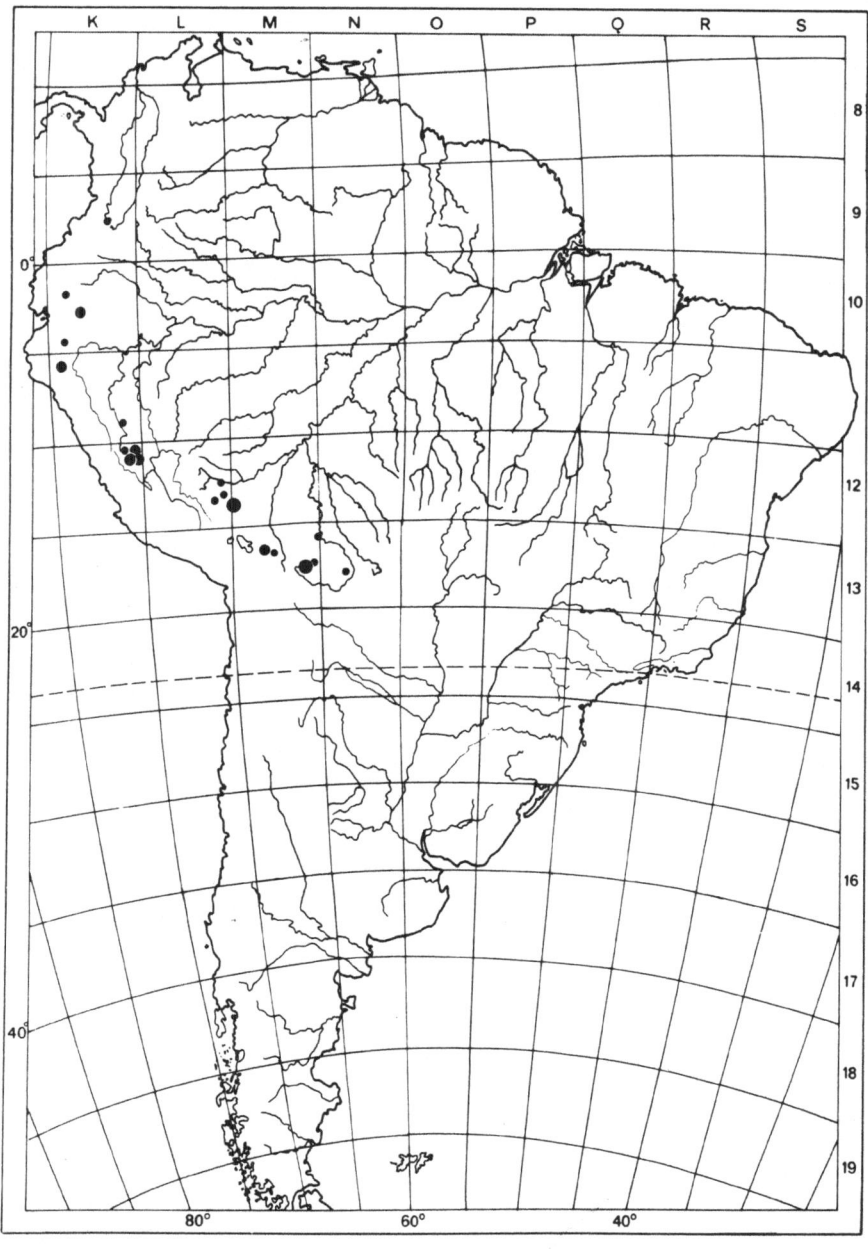

Fig. 55. Range of *Euryglottis guttiventris* R. & J.

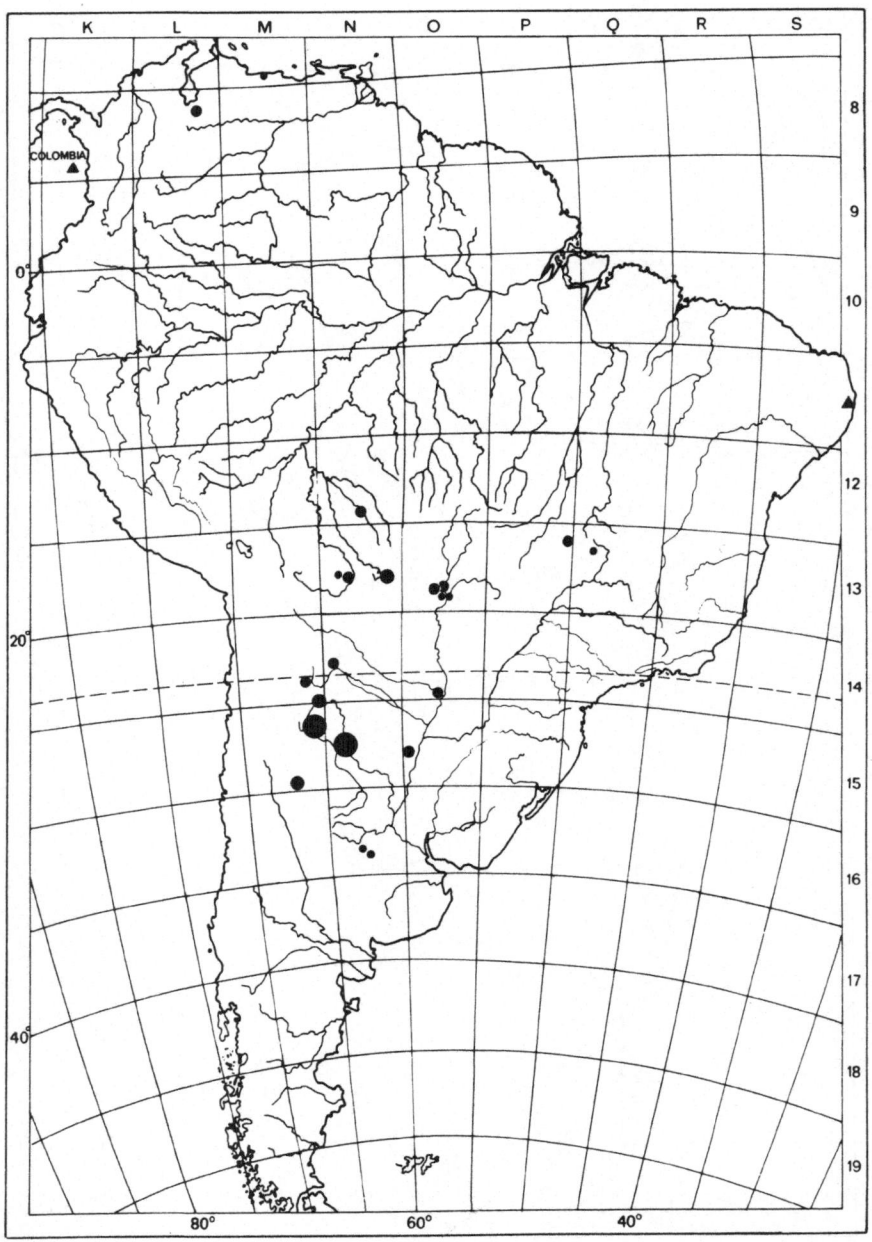

Fig. 56. Ranges of *Callionima grisescens* (ROTHSCH.) ●, *Callionima grisescens elegans* GEH-LEN ▲.

164

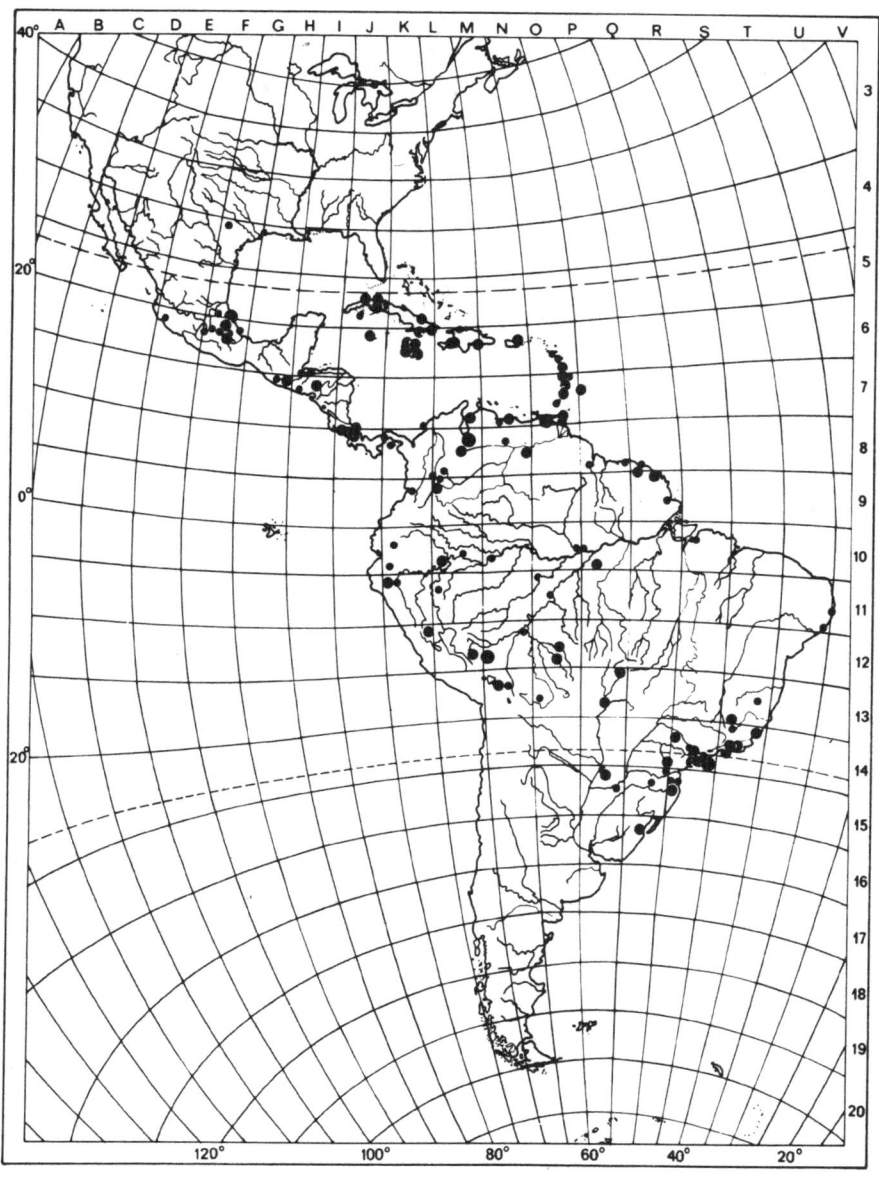

Fig. 57. Range of *Protambulyx strigilis* (L.)

165

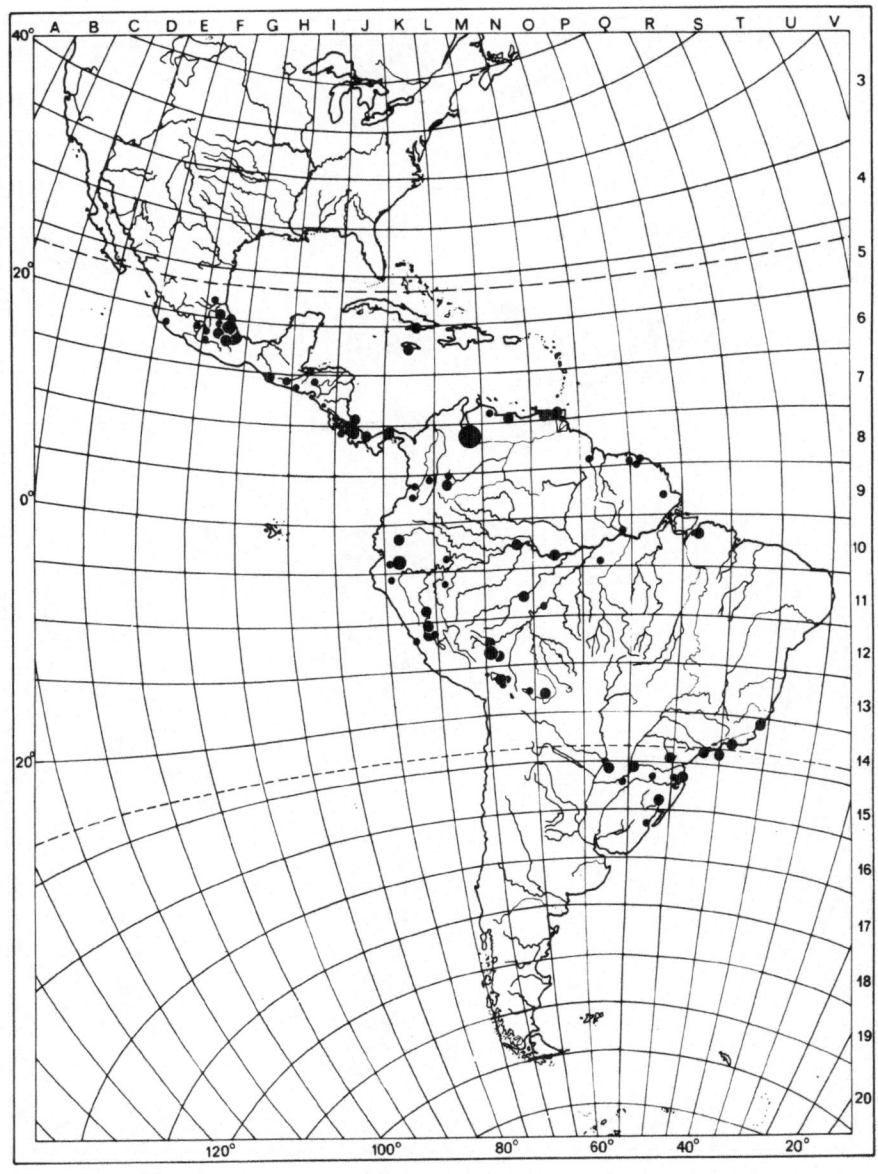

Fig. 58. Range of *Amplypterus gannascus* (STOLL).

166

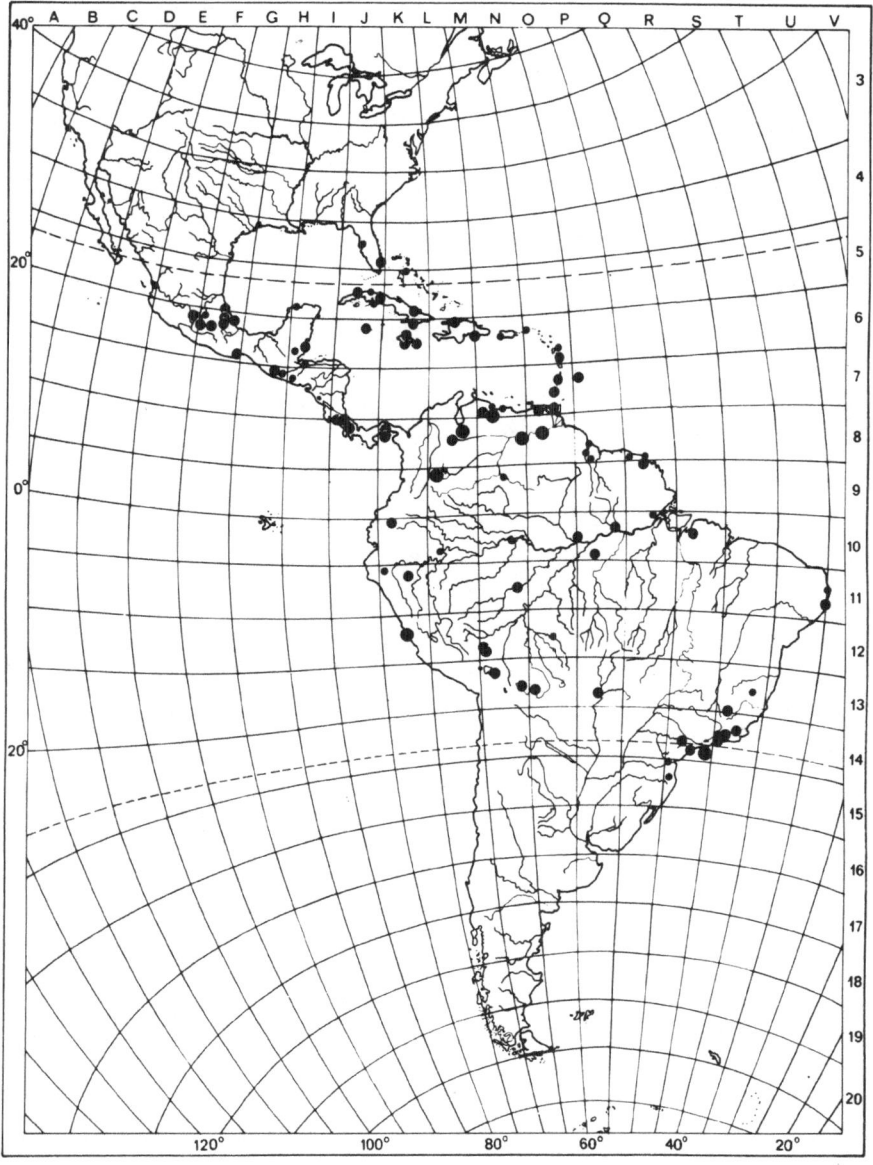

Fig. 59. Range of *Pseudosphinx tetrio* (L.).

167

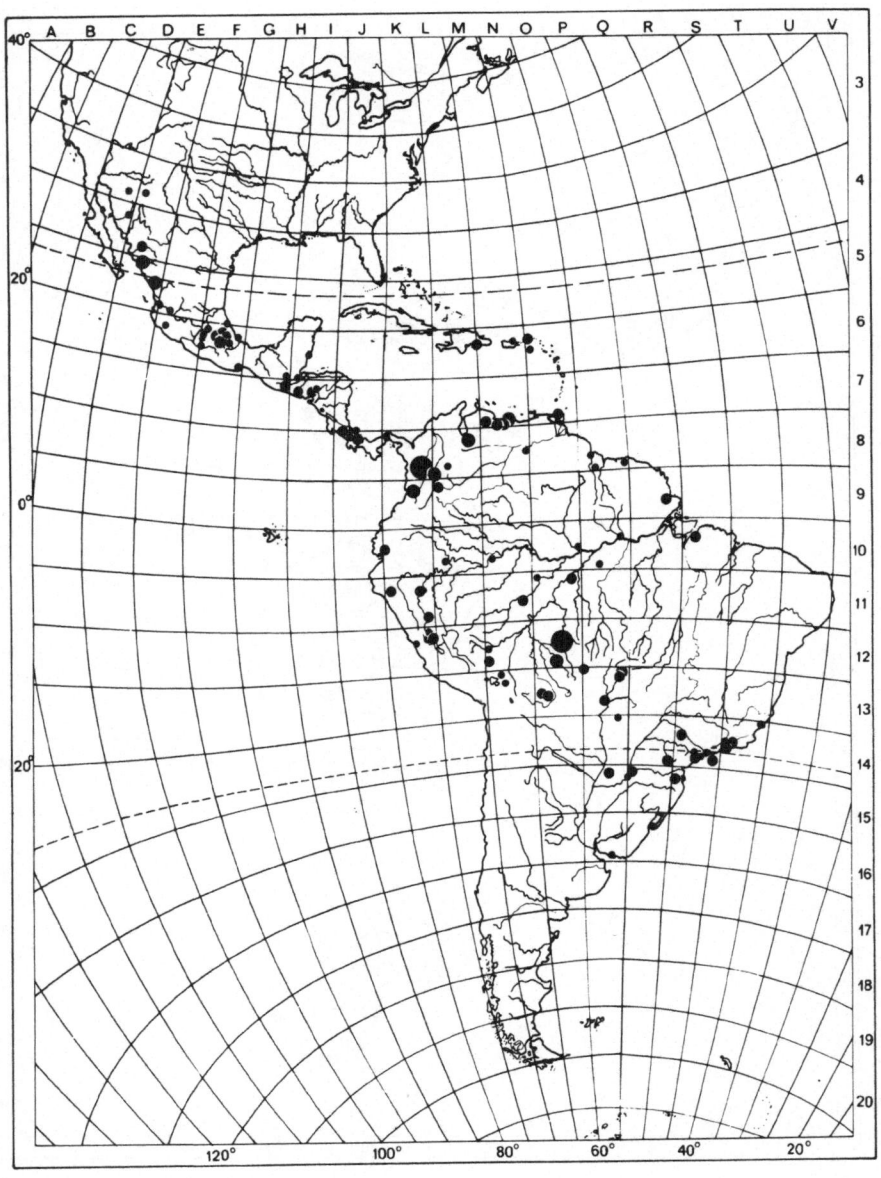

Fig. 60. Range of *Callionima parce* (F.).

168

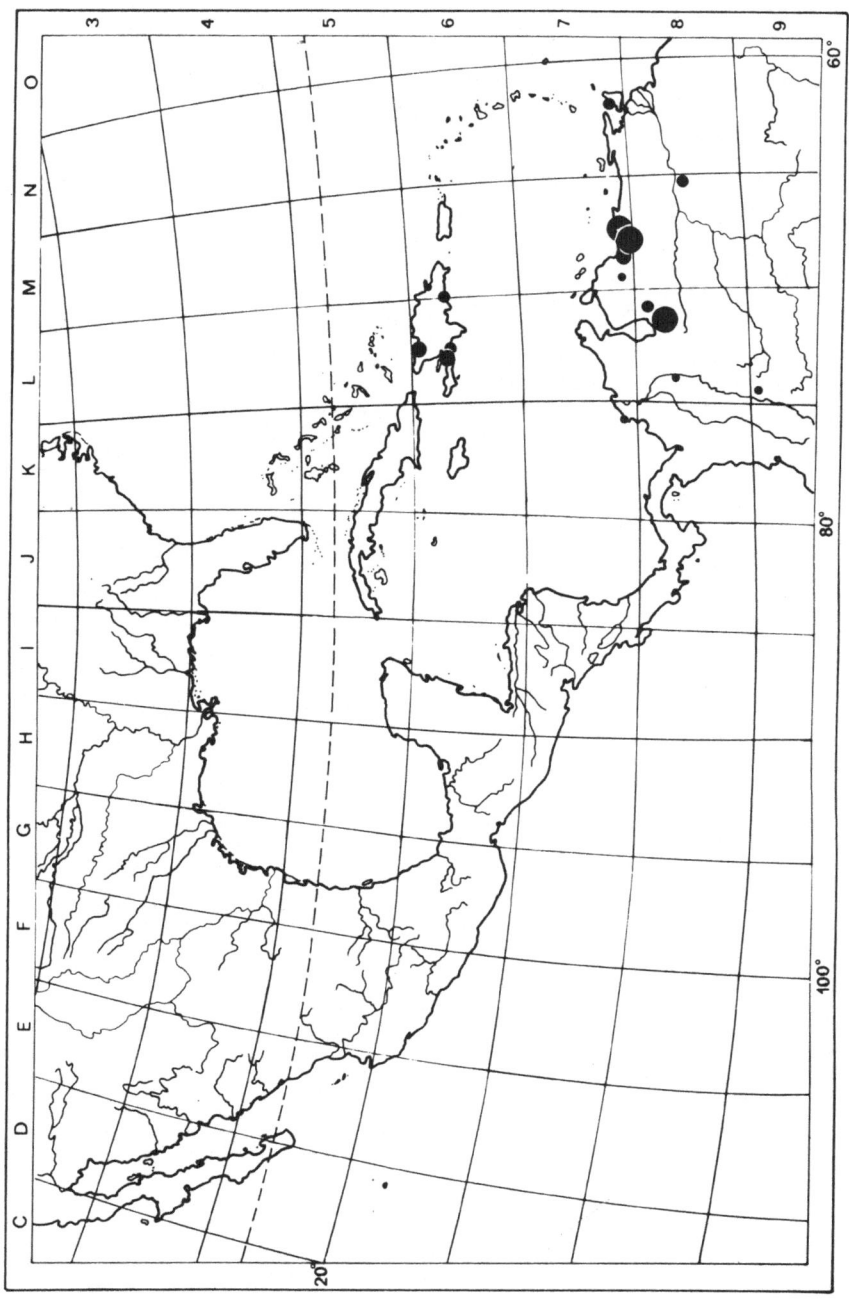

Fig. 61. Ranges of *Callionima calliomenae* (SCHAUF.).

169

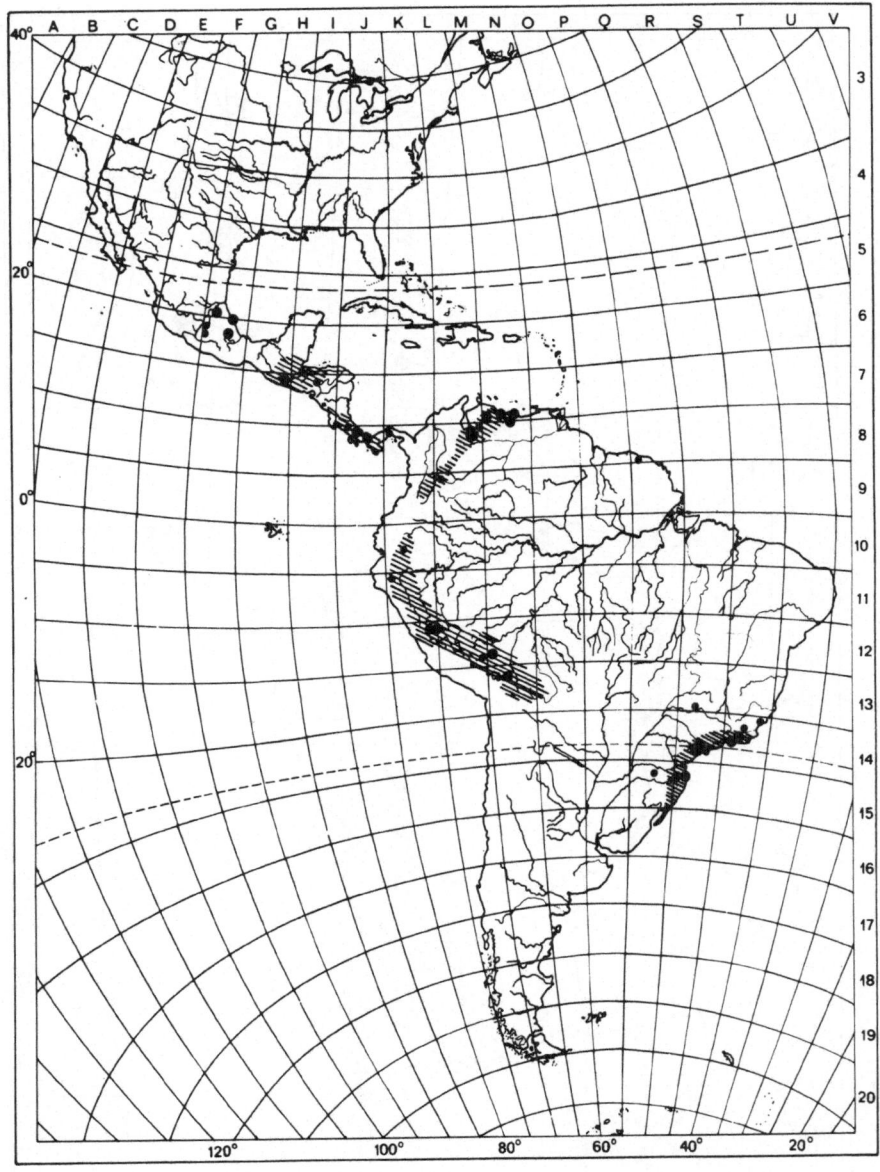

Fig. 62. Disjunct distribution of *Perigonia stulta* H.-S. and *Xylophanes titana* (DRC.) ● between the Andes and the Serra do Mar.

170

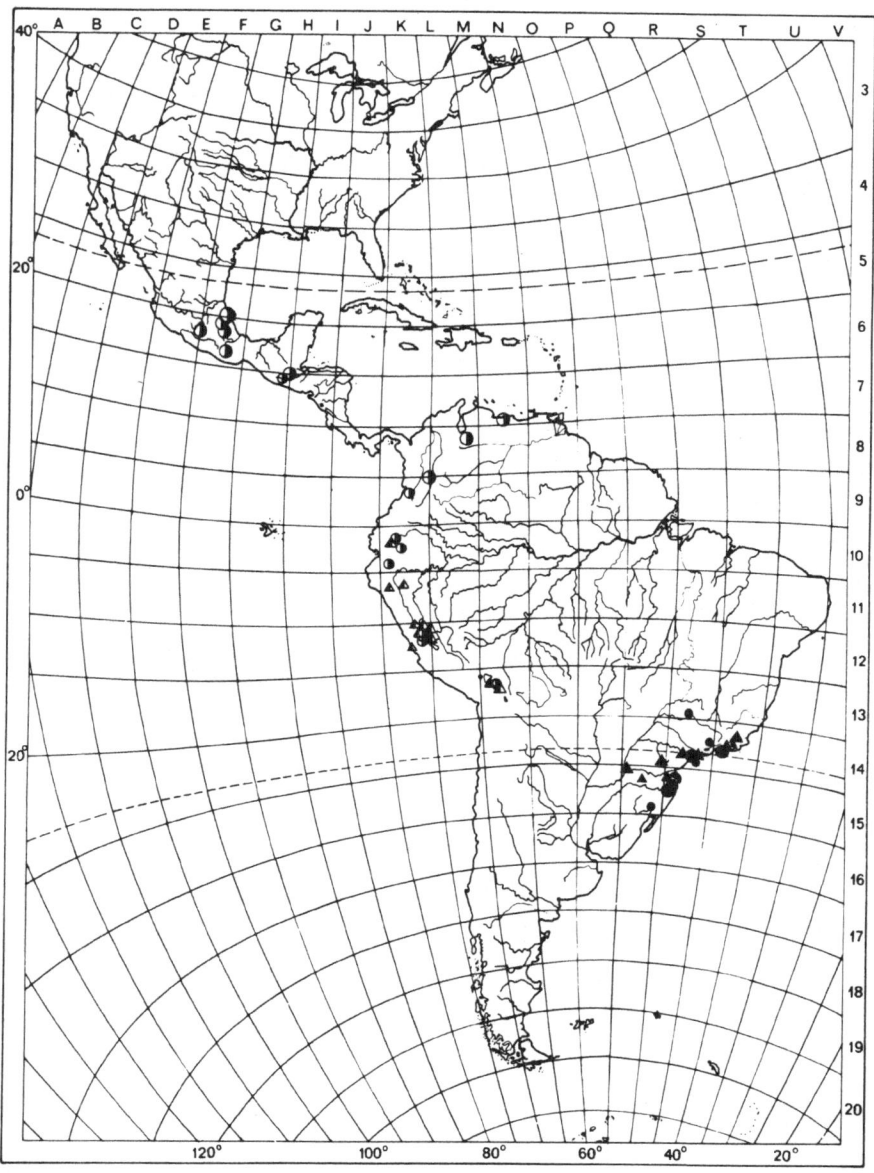

Fig. 63. Differentiation of subspecies of *Manduca pellenia* (H-S.) and *Nyceryx nictitans* (BDV.) between the Serra do Mar and the Andean-Central American area: *Manduca pellenia pellenia* (H.-S.) ⊙, *Manduca pellenia janira* (JORD.) ●. *Nyceryx nictitans nictitans* (BDV.) ▲. *Nyceryx nictitans saturata* R. & J. △.

171

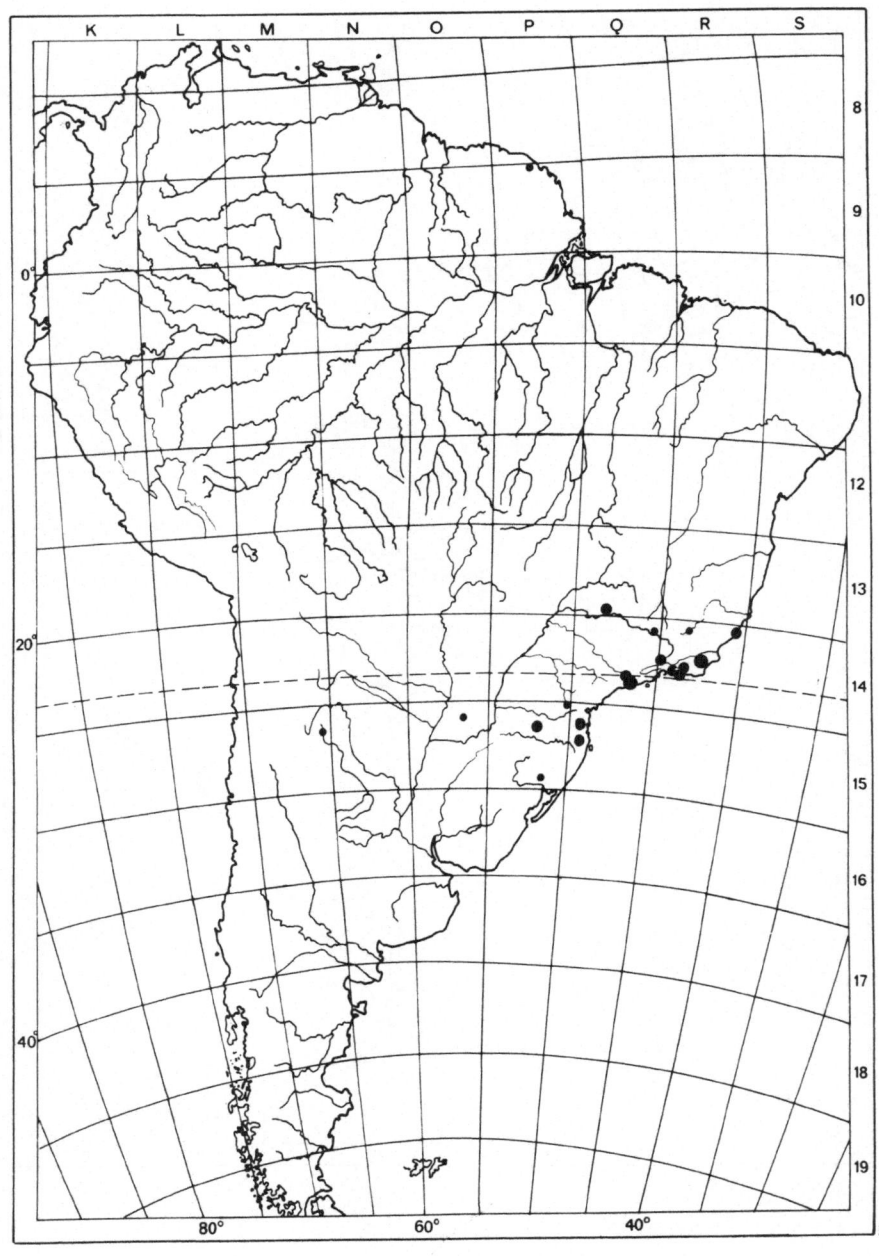

Fig. 64. Range of *Amplypterus eurysthenes* (FLDR.).

172

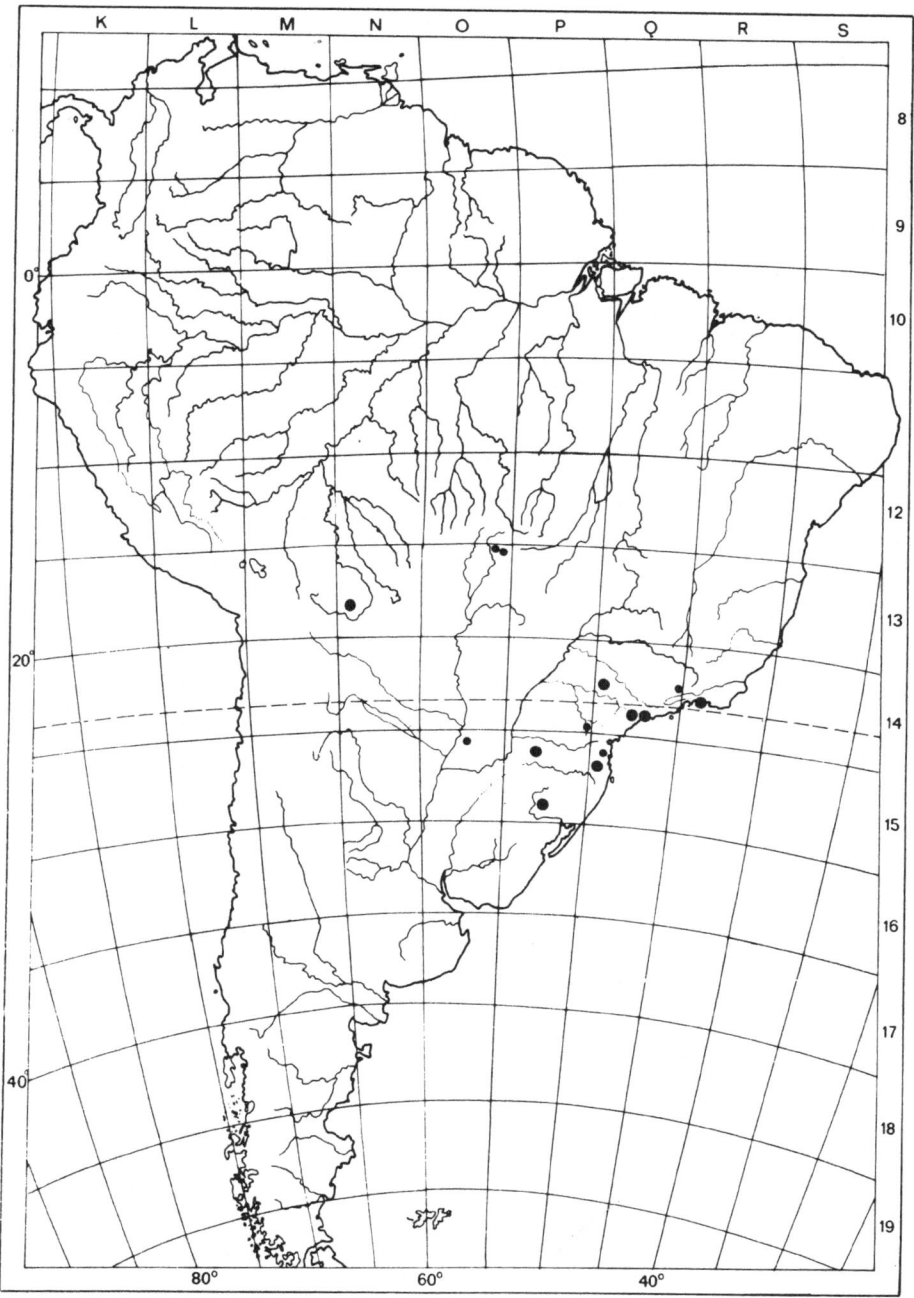

Fig. 65. Range of *Manduca incisa* (WLK.).

173

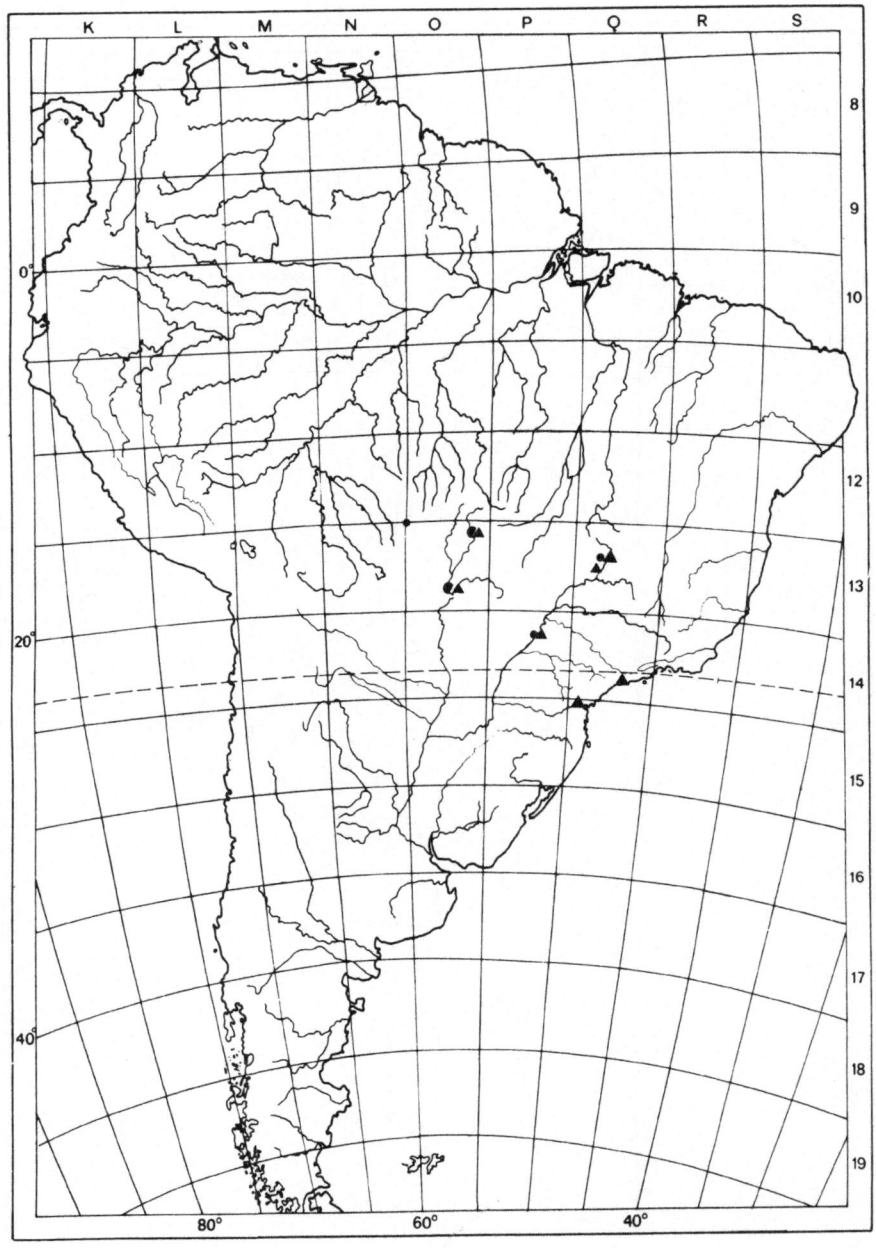

Fig. 66. Ranges of *Manduca corumbensis* (CLARK) ● and *Neogene curitiba* JONES ▲.

174

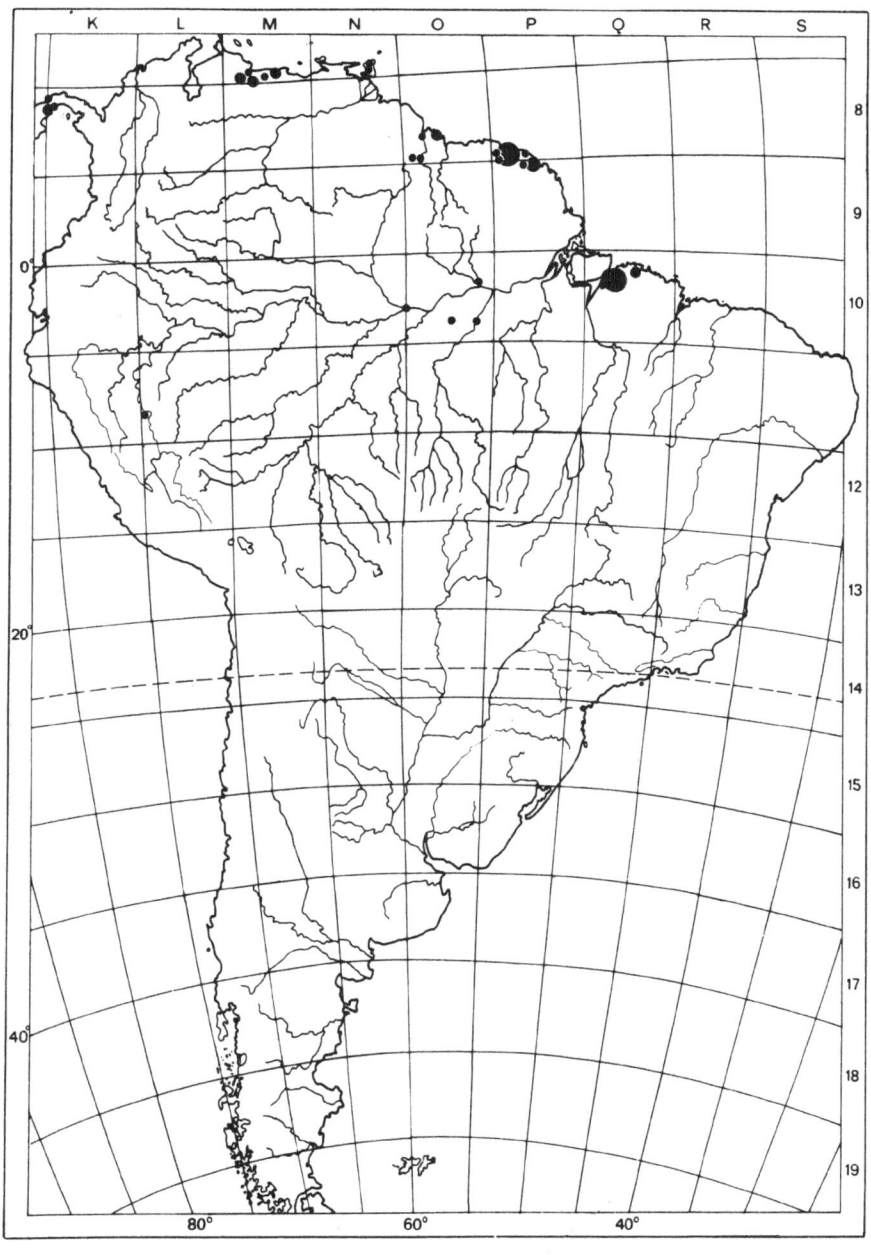

Fig. 67. Range of *Isognathus scyron* (STOLL).

175

b. Locality index

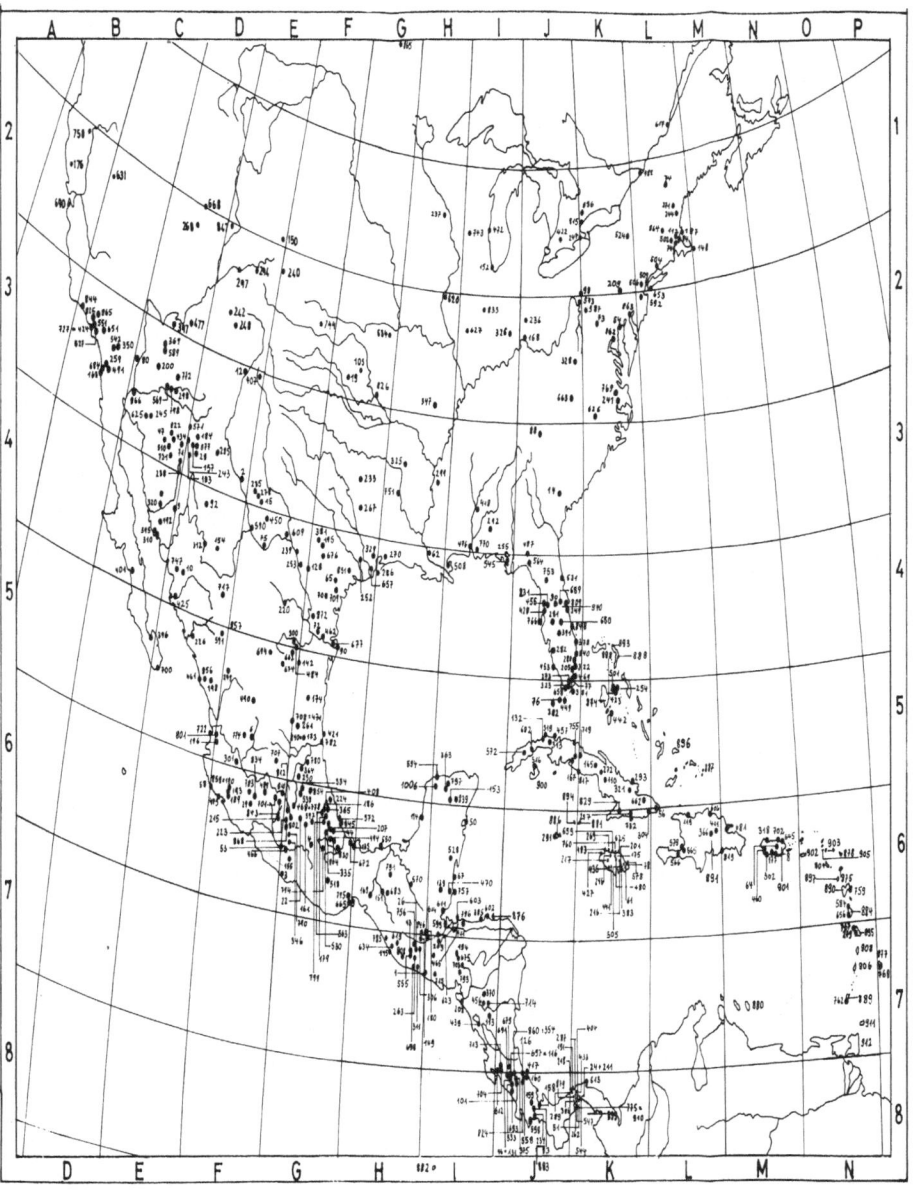

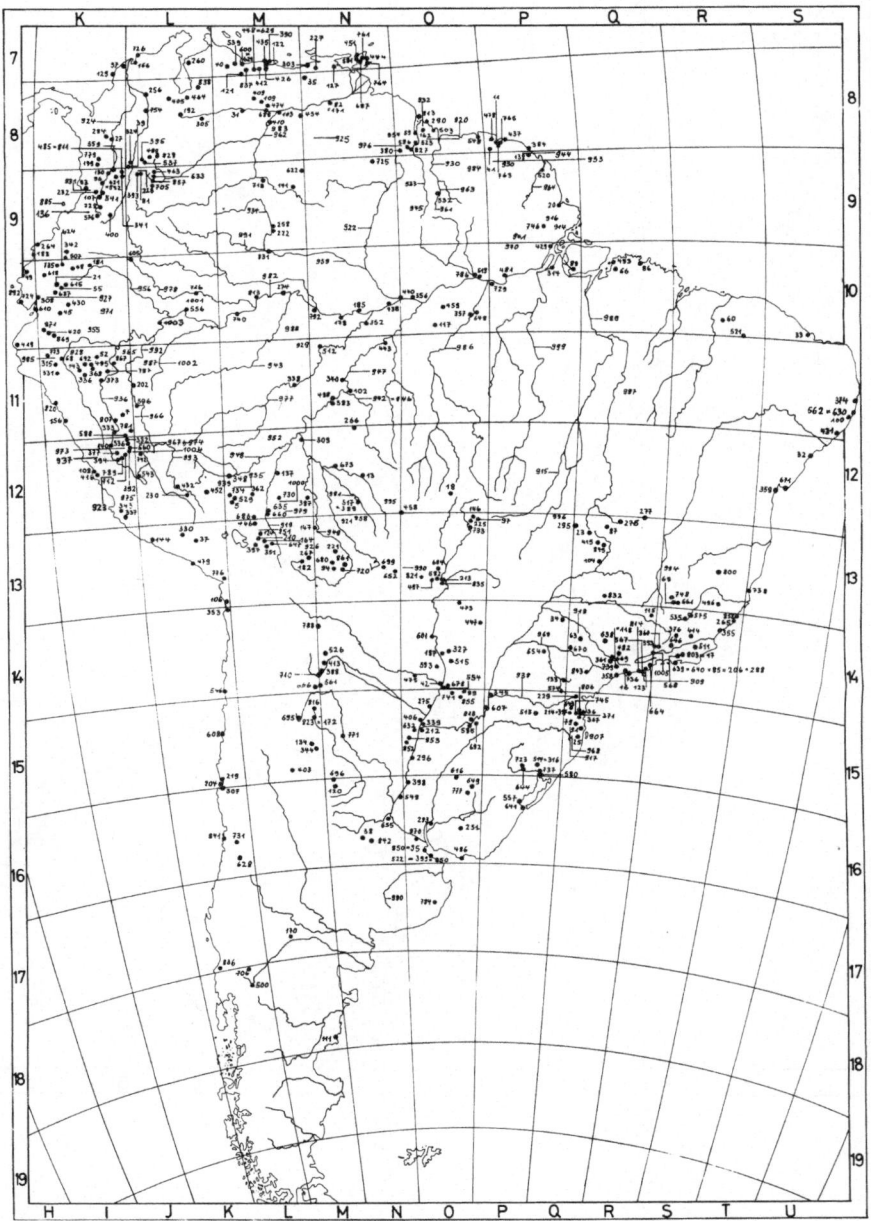

51	Balboa, Panama Can. Zone	K 8
52	Balsapuerto, Peru	K 11
53	Balsas, Guerrero, Mexico	F/G 6
54	Baltimore, Maryland, USA	K 2
55	Baños, Ecuador	K 10
56	Baracoa, Cuba	L5/6
57	Baranquilla, Colombia	K/L 7
58	Barra de Navidad, Jalisco, Mexico	D/E 6
59	Bartica, British Guiana	O 8
60	Batalha, Piauí, Brazil	R 10
61	Bath, Jamaica	K 6
62	Baton Rouge, Louisiana, USA	H 4
63	Bauru, São Paulo, Brazil	Q 14
64	Bayamón, Puerto Rico	M 6
65	Beeville, Texas, USA	G 4
66	Belém = Pará, Pará, Brazil	Q 10
67	Belize, British Honduras	I 9
68	Bellavista, Rio Marañon, Catamarca, Peru	K 11
69	Belo Horizonte, Minas Gerais, Brazil	R 13
70	Benavides, Texas, USA	G 4
71	Benson, Arizona, USA	D/E 3
72	Bentson, Rio Grande Valley State Park, Texas, USA	G 4
73	Berkeley Springs, W. Virginia, USA	K 2
74	Bethlehem, New Hampshire, USA	L 1
75	Big Bend National Park, Texas, USA	F 4
76	Big Pine, Key Island, Florida, USA	J 5
77	Biscayne Bay, Florida, USA	J 4
78	Blue Mountain Pk., Jamaica	K 6
79	Blumenau, Santa Catarina, Brazil	Q 15
80	Blythe, Colorado Desert, Calif., USA	D 3
81	Bogotá, Colombia	L 9
82	Bolivar (= Ciudad de B.), Venezuela	N 8
83	Boquete, Chiriquí, Panama	J 8
84	Boston, Mass., USA	L 1
85	Botafogo (= part of City of Rio de Janeiro), Brazil	R 14
86	Bragança, Pará, Brazil	Q 10
87	Brasilia, Brazil	Q 13
88	Brevard, N. Carolina, USA	J 3
89	Breves, Ilha de Marajó, Pará, Brazil	P 10
90	Brownsville, Texas, USA	G 4
91	Brusque, Santa Catarina, Brazil	Q 15
92	Buenaventura, Chihuahua, Mexico	E 4
93	Buenaventura, Colombia	K9
94	Buenavista, Bolivia	N 13
95	Buenos Aires, Argentina	O 16
96	Buga, Colombia	K 9
97	Burity, 30 mi NE of Cuiabá, Mato Grosso, Brazil	G 8
98	Butler, Penna., USA	J/K 1
99	Caaguazu, Paraguay	O 15
100	Cabo near Recife, Pernambuco, Brazil	T 11
101	Cachi, Costa Rica	J 8
102	Calama, Rio Madeira, Rondônia, Brazil	N 11
103	Caicara, Venezuela	M 8
104	Caldas Novas, Goias, Brazil	Q 13

105	Caldwell, Kansas, USA	G 2
106	Caleta Buena, Chile	L/M 13
107	Cali, Cauca, Colombia	K 9
108	Callao, Peru	K 12
109	Camaguan, Guárico, Venezuela	M 8
110	Camagüey, Cuba	K 5
111	Camarones, Chubut, Argentina	M 18
112	Cambridge Mass., USA	L 1
113	Camoapa, Nicaragua	I 7
114	Campeche, Mexico	H 6
115	Campo Belo, Minas Gerais, Brazil	Q 14
116	Candelaria Mts., south of San José, Costa Rica	J 8
117	Cantagalo, Amazonas, Brazil	O 10
118	Cantareira Mt. near São Paulo, Brazil	Q 14
119	Cap Haitien, Haiti	L 6
120	Capilla del Monte, Cordoba, Argentina	N 16
121	Carabobo, Venezuela	M 7/8
122	Caracas, Venezuela	M 7
123	Caraguatatuba, São Paulo, Brazil	Q 14
124	Caranavi, Rio Songo, Bolivia	M 13
125	(number not used by mistake)	
126	Carillo, Costa Rica	J 7/8
127	Caripito, Venezuela	M 7
128	Carrizo Springs, Texas, USA	F/G 4
129	Cartagena, Colombia	K 7
130	Cartago, Colombia	K 9
131	Cartago, Costa Rica	J 8
132	Casablanca near Havana, Cuba	J 5
133	Castro, Paraná, Brazil	P 14
134	Catamarca, Catamarca, Argentina	M 15
135	Catemaco, Vera Cruz, Mexico	G/H 6
136	Cauca, Prov. of, Colombia	K 9
137	Cavinas, Bolivia	M 12
138	Cayenne, French Guiana	P 9
139	Cayo, Brit. Honduras	I 6
140	Cerro de Pasco, Peru	K 12
141	Cerro Duido, Venezuela	M 9
142	Cerro Potosí, Nuevo Leon, Mexico	F/G 5
143	Chachapoyas, Peru	K 11
144	Chala, Peru	L 13
145	Champerico, Guatemala	H 7
	Chanchamayo see no. 1004	
146	Chapada near Cuiabá, Mato Grosso, Brazil	O 13
147	Charaplaya, Bolivia	M/N 13
148	Chatham, Mass., USA	L/M 1
149	Cheje, south of Vera Paz, Guatemala	H 7
150	Cheyenne, Wyoming, USA	E 1
151	Chiapa del Corzo, Chiapas, Mexico	H 6
152	Chicago, Illinois, USA	I 1
153	Chichen Itza, Yucatán, Mexico	I 5
154	Chihuahua, Chihuahua, Mexico	E 4
155	Chilpancingo de los Bravos, Guerrero, Mexico	G 6

209	Cornwall, Penna., USA	K 1/2
210	Coroico, Bolivia	M 13
211	Corozal, Panama Can. Zone	K 8
212	Corrientes, Corrientes, Argentina	O 15
213	Corumba, Mato Grosso, Brazil	O 13
214	Corupa, Santa Catarina, Brazil	Q 15
215	Corupo, Mexico	F 6
216	Craighton, St. Andrew, Jamaica	K 6
217	Christiana, Jamaica	K 6
218	Cristobal, Panama Can. Zone	K 8
	Cruzeiro see no. 1005	
219	Cruz Grande, Coquimbo, Chile	L 15
220	Cuatrocienegas, Mexico	F 4
221	Cuatro Ojos, Bolivia	N 13
222	Cucui, Amazonas, Brazil	M 9
223	Cuernavaca, Mexico	G 6
224	Cuesta de Misantla (= Misantla), Vera Cruz, Mexico	G 6
225	Cuiabá (= Cuyabá), Mato Grosso, Brazil	O 13
226	Culiacán, Sinaloa, Mexico	E 5
227	Cumanacoa (= Mirasol), Venezuela	N 7
228	Cundinamarca Div., Colombia	L 9
229	Curitiba, Paraná, Brazil	Q 15
230	Cuzco, Peru	L 12
231	Dade City, Florida, USA	J 4
232	Dagua, Colombia	K 9
233	Dallas, Texas, USA	G 3
234	David, Panama	J 8
235	Davis Mts., Texas, USA	F 3
236	Dayton, Ohio, USA	J 2
237	Decorah, Iowa, USA	H 1
238	Del Rio, Mexico	D 3
239	Del Rio, Texas, USA	D/E 3
	Demerara cf. Georgetown, Brit. Guiana	O 8
240	Denver, Colorado, USA	E/F 2
241	Dismal Swamp, Virginia/N. Carolina, USA	K 2
242	Dolores, Colorado, USA	D 2
243	Douglas, Arizona, USA	E 3
244	Dover, New Hampshire, USA	L 1
245	Dripping Springs, Organ Pipe National Monument, Arizona, USA	C 3
246	Duncans, Jamaica	K 6
247	Dunnville, Ontario, Canada	K 1
248	Durango, Colorado, USA	E 2
249	Durango, Durango, Mexico	E/F 5
250	Durango, Hidalgo, Mexico	G 5
251	Durazno, Uruguay	O 16
252	Eagle Lake, Texas, USA	G 4
253	Eagle Pass, Texas, USA	F 4
254	East End Point, New Providence I., Bahamas	K 5
255	Ebro, Florida, USA	I 3/4
256	El Banco, Magdalena Valley, Colombia	L 8
257	El Caney, Cuba	K 5/6
258	El Carmen, Venezuela	M 9
259	El Cayon, Calif., USA	C 3

260	El Mene, Venezuela	L 7
261	El Salto, San Luis Potosí, Mexico	G 5
262	Empire, Panama Can. Zone	J/K 8
263	Escuintla, Guatemala	H 7
264	Esmeraldas, Ecuador	K 9
265	Espirito Santo, Espirito Santo, Brazil	R 14
266	Espirito Santo, Rondônia, Brazil	N 12
267	Espiritu Santo, Cochabamba, Bolivia	M 13
268	Eureka, Utah, USA	D 2
269	Falmouth, Jamaica	K 6
270	Fannett, Texas, USA	H 4
271	Farmington, New Hampshire, USA	L 1
272	Florida, Cuba	K 5
273	Florida City, Florida, USA	J 4
274	Fonte Boa, Amazonas, Brazil	M 10
275	Formosa, Argentina	O 15
276	Formosa, Brazil	Q 13
277	Formoso, Brazil	Q 12/13
278	Fort Davis, Texas, USA	F 3
279	Fort de France, Martinique, Lesser Antilles	N 7
280	Fort Lauderdale, Florida, USA	J 4
281	Fort Meade, Florida, USA	J 4
282	Fort Myers, Florida, USA	J 4
283	Fray Bentos, Uruguay	O 16
284	Frontino, Colombia	K 8
285	Gage, New Mexico, USA	E 3
286	Galveston, Texas, USA	H 4
287	Gamboa, Panama Can. Zone	K 8
288	Gavéa (= part of City of Rio de Jan.) Brazil	R 14
289	Gatun, Panama Can. Zone	K 8
290	Georgetown (= Demerara), Brit. Guiana	O 8
291	Georgetown, Grand Cayman I.	J 6
292	Georgiana, Alabama, USA	I 3
293	Gibara, Cuba	K 5
294	Glenwood Springs, Colorado, USA	E 2
295	Goias, Goias, Brazil	P/Q 13
296	Goya, Corrientes, Argentina	O 15
297	Grand Junction, Colorado, USA	E 2
298	Granite Reef near Tempe, Arizona, USA	D 3
299	Greenville, Mississippi, USA	H 3
300	Grutas Carcia near Monterrey, N. Leon, Mexico	F 4/5
301	Guadalajara, Jalisco, Mexico	F 5/6
302	Guánica, Puerto Rico	M 6
303	Guanta, Venezuela	N 7
304	Guantanamo, Cuba	K 5/6
305	Guasdualito, Venezuela	L 8
306	Guatemala City, Guatemala	H 7
307	Guayacan, Chile	L 15/16
308	Guayaquil, Ecuador	K 10
309	Guayavamerin, Bolivia	M 12
310	Guaymas, Sonora, Mexico	D 4
311	Guazacapan, Guatemala	H 7
312	Guerrero, Hidalgo, Mexico	E 4

313	Guines, Cuba	J 5
314	Gurupa, Pará, Brazil	P 10
315	Hacienda Taulis, Peru	K 11
316	Hamburgo Velho (= old part of Novo Hamburgo), Rio Grande do Sul, Brazil	P 15
317	Hansa Humboldt (= Corupa since 1958; previously part of Jaraguá do Sul), S. Catarina, Brazil	Q 15
318	Hatillo, Puerto Rico	M 6
319	Havana (= Habana), Cuba	J 5
320	Hermosillo, Mexico	D 4
321	Holguin, Cuba	K 5
322	Hollywood, Florida, USA	J 4
323	Homestead, Florida, USA	J 4/5
324	Honda, Colombia	L 8/9
325	Hope, Arkansas, USA	H 3
326	Hope, Indiana, USA	I 2
327	Horqueta, Paraguay	O 14/15
328	Hot Springs, Virginia, USA	J/K 3
329	Houston, Texas, USA	G/H 4
330	Huambo, Peru	L 13
331	Huambos, Peru	K 11
332	Huancabamba, Peru	K 12
333	Huánuco, Peru	K 11/12
334	Huasca, Mexico	G 6
335	Huatuxco, Vera Cruz, Mexico	G 6
336	Huayabamba, Peru	K 11
337	Huayuri (= Pampa de H.), Peru	K 12
338	Huitanão, Rio Purus, Amazonas, Brazil	M 11
339	Humaita, Paraguay	O 15
340	Humaita, Rio Madeira, Brazil	N 11
341	Ibagué, Colombia	K 9
342	Ibarra, Ecuador	K 9
343	Ica, Peru	K 12
344	Icaño, Catamarca, Argentina	M 15
345	Iguaçu (= Foz, do I.), Paraná, Brazil	O/P 15
346	Iguala, Mexico	G 6
347	Imboden, Arkansas, USA	H 2
348	Inambari, Carabaya, Peru	M 12
349	Indian River City, Florida, USA	J 4
350	Indio, Calif., USA	C 3
351	Inquisivi, Bolivia	M 13
352	Ipiranga, Rio Purus, Amazonas, Brazil	N 10
353	Iquique, Chile	L/M 14
	Iquitos see no. 1003	
354	Irazu (volcano), Costa Rica	J 7/8
355	Itabapoana, Rio de Janeiro, Brazil	R 14
356	Itacoatiara, Amazonas, Brazil	O 10
357	Itaituba, Rio Tapajós, Pará, Brazil	O 10
358	Itanhaém, São Paulo, Brazil	Q 14
359	Itaparica, Bahía, Brazil	S 12
360	Itatiaia Mt. (= Agulhas Negras), Minas Gerais, Brazil	Q/R 14
361	Itu, São Paulo, Brazil	Q 14
362	Ixiamas, Bolivia	M 12

363	Izamal, Yucatán, Mexico	I 5
364	Jacala, Hidalgo, Mexico	G 5
365	Jalapa, Vera Cruz, Mexico	G 6
366	Jaracaboá, Dominican Republic	L 6
367	Jaragua do Sul, Santa Catarina, Brazil	Q 15
368	Jepelacio, N. Peru	K 11
369	Jerome, Arizona, USA	D 3
370	Jinotega, Nicaragua	I 7
371	Joinville, Santa Catarina, Brazil	Q 15
372	José Cardel, Vera Cruz, Mexico	G 7
373	Juanjui, Peru	K 11
374	Juan Pessoa (= João Pessoa), Paraiba, Brazil	T 11
375	Juan Vinas, Costa Rica	J 8
376	Juiz de Fora, Minas Gerais, Brazil	R 14
377	Junin, Peru	K 12
378	Jupiter, Florida, USA	J 4
379	Kaibab National Forest (near Grand Canyon), Arizona, USA	D 2
380	Kaieteur Falls, Brit. Guiana	N/O 8
381	Kerrville, Texas, USA	G 4
	Key Largo see Largo Key, Florida, USA	J 5
382	Key West, Florida, USA	J 5
383	Kingston, Jamaica	K 6
384	Kourou, French Guiana	P 8/9
385	La Ceiba, Honduras	I 6
386	La Chorrera, Panama	K 8
387	La Esperanza, Bolivia	M 12
388	La Esperanza, Jujuy, Argentina	N 14
389	La Esperanza (Nuflo de Chavez), E. Bolivia	K 12
390	La Guaira, Venezuela	M 7
391	Lake Placid, Florida, USA	J 4
392	La Merced, Chanchamayo, Peru (cf. no. 1004)	K 12
393	La Mesa, Colombia	L 9
394	La Oroya, Peru	K 12
395	La Palma, Cundinamarca, Colombia	L 8
396	La Paz, Baja California Sur, Mexico	D 5
397	La Paz, Bolivia	M 13
398	La Paz, Entre Rios, Argentina	O 16
399	La Plata, Argentina	O 16/17
400	La Plata, Colombia	K 9
401	La Purisima, Baja California Sur, Mexico	D 4
402	Largo Key, Florida, USA	J 5
403	La Rioja, La Rioja, Argentina	M 15
404	Las Cascadas, Panama Can. Zone	J/K 8
405	Las Cruces, Venezuela	L 8
406	Las Palmas, Chaco, Argentina	O 15
407	Las Vegas, New Mexico, USA	E 3
408	Las Vigas, Vera Cruz, Mexico	G 6
409	La Union, Venezuela	M 8
410	La Urbana, Bolivar, Venezuela	M 8
411	La Vega, Dominican Republic	L 6
412	La Victoria, Venezuela	M 7
413	Ledesma (= Libertador General San Martin), Jujuy, Argentina	N 15
414	Leopoldina, Rio de Janeiro, Brazil	R 14

467	Mexia, Texas, USA	G 3
468	Mexico City, Mexico	G 6
469	Miami, Florida, USA	J 4
470	Middlesex, British Honduras	I 6
471	Miguel Hidalgo, Tamaulipas, Mexico	F/G 5
472	Milwaukee, Wisconsin, USA	I 1
473	Miranda, Mato Grosso, Brazil	O 14
474	Miranda (= Puerto M.), Venezuela	M 8
	Misantla see Cuesta de M., Vera Cruz, Mexico	G 6
475	Mision F. Tacaaglé, Formosa, Argentina	O 14
476	Mobile, Alabama, USA	I 3
477	Moenave Reservoir, Indian Village near Oraibi, Arizona, USA	D 2
478	Moengo, Surinam	P 8
479	Mollendo, Peru	L 13
480	Moneague, Jamaica	K 6
481	Monte Alegre, Lower Amazon, Brazil	P 10
482	Monte Alegro, São Paulo, Brazil	Q 14
483	Montego Bay, Jamaica	K 6
484	Monterrey, Mexico	F 4/5
485	Monte Tolima (=Nevado de T.), Colombia	K 9
486	Montevideo, Uruguay	O 16/17
487	Monticello, Florida, USA	J 3/4
488	Montreal, Canada	L 1
489	Morelia, Michoacan, Mexico	F 6
490	Morelos, Zacatecas, Mexico	F 5
491	Morena Reservoir, Lake Morena, Cleveland National Forest, Calif., USA	C 3
492	Moreno, Sonora, Mexico	D 4
493	Mosqueiro, Rio do Pará, Brazil	Q 10
494	Motzorongo, Vera Cruz, Mexico	G 6
495	Moyobamba, Peru	K 11
496	Mutum, Minas Gerais, Brazil	R 13
497	Mutum near Puerto Suarez, Bolivia	O 13
498	Mutum, Rondônia, Brazil	N 11
499	Muzo, Colombia	L 8
500	Nahuel Huapi, Rio Negro, Argentina	L 18
501	Nassau, New Providence I., Bahamas	K 5
502	Natrick, Massachusetts, USA	L 1
503	New Amsterdam, British Guiana	O 8
504	Newark, New Jersey, USA	L 1
505	Newcastle, Jamaica	K 6
506	New Haven, Connecticut, USA	L 1
507	Nono, Ecuador	K 9/10
508	New Orleans, Louisiana, USA	H/I 4
509	New York, New York, USA	L 1
510	Nogales, Arizona, USA	D 3
511	Nova Friburgo, Rio de Janeiro, Brazil	R 14
512	Nova Olinda, Rio Purus, Amazonas, Brazil	N 11
513	Nova Teutonia, Rio Grande do Sul, Brazil	P 15
514	Novo Hamburgo, Rio Grande do Sul, Brazil	P 15
515	Nueva Germania, Paraguay	O 14
516	Nueva Gerona, Isla de Pinos, Cuba	J 5
517	Nuflo de Chavez (La Esperanza), E. Bolivia	N 8

518	Oaxaca, Mexico	G 6
519	Obidos, Pará, Brazil	O 10
520	Oiapoque, French Guiana	P 9
521	Oiticica, Ceara, Brazil	R 10/11
522	Olivos (= part of Buenos Aires), Argentina	O 16/17
523	Omai, British Guiana	O 8
524	Oneonta, New York, USA	K 1
525	Oracabessa, Jamaica	K 6
526	Oran, Salta, Argentina	N 14
527	Orange, California, USA	C 3
528	Orange Walk, British Honduras	I 6
529	Oraya, Carabaya, Peru	M 12
	Organ Pipe National Monument see Dripping Springs, Arizona, USA	C 3
530	Orizaba, Vera Cruz, Mexico	G 6
531	Ormond, Florida, USA	J 4
532	Oronoque, British Guiana	O 9
533	Orosi, Costa Rica	J 8
534	Ottawa, Kansas, USA	G 2
535	Ouro Preto, Minas Gerais, Brazil	R 14
536	Oxapampa, Peru	K 12
537	Pacho, Colombia	L 8/9
538	Pachuca, Mexico	G 6
539	Palma Sola, Venezuela	M 7
540	Palm Beach, Florida, USA	J 4
541	Palmira, Cauca, Colombia	K 9
542	Palm Springs, Calif., USA	C 3
543	Pampas, Peru	K/L 12
544	Panama, Panama	K 8
545	Panama City, Florida, USA	I 4
546	Paposo, Chile	L 14
547	Paraiso, Panama Can. Zone	K 8
548	Paramaribo, Surinam	O 8
549	Parana, Entre Rios, Argentina	N/O 16
550	Pareiso, Vera Cruz, Mexico	H 6
551	Pasadena, Calif., USA	C 3
552	Paso de Quindio, Colombia	K 9
553	Passa Quatro, Minas Gerais, Brazil	Q/R 14
554	Patino-cué, Paraguay	O 15
555	Patulul, Guatemala	H 7
556	Pebas, Amazon, Peru	L 10
557	Pelotas, Rio Grande do Sul, Brazil	O 16
558	Peralta, Costa Rica	J 8
559	Pereira, Colombia	K 9
560	Perene, Peru	K 12
561	Perico, Jujuy, Argentina	M/N 14
562	Pernambuco (= Recife), Pernambuco, Brazil	T 11
563	Perote, Mexico	G 6
564	Perry, Florida, USA	J 4
565	Petion-Ville, Haiti	L 6
566	Petit Goáve, Haiti	L 6
567	Petrobas, São Paulo, Brazil	Q 14
568	Petrópolis, Rio de Janeiro, Brazil	R 14
569	Phoenix, Arizona, USA	D 3

570	Piedras Negras, Guatemala	H 6
571	Pinalenon Mts., Arizona, USA	E 3
572	Pinar del Rio, Cuba	J 5
573	Pittsburgh, Penna., USA	J/K 2
574	Ponta Grossa, Paraná, Brazil	P 14/15
575	Ponte Nova, Minas Gerais, Brazil	R 14
576	Popayan, Colombia	K 9
577	Portal, Arizona, USA	E 3
578	Port Antonio, Portland, Jamaica	K 6
579	Port-au-Prince, Haiti	L 6
580	Porto Alegre, Rio Grande do Sul, Brazil	P 15/16
581	Port of Spain, Trinidad	N 7
582	Porto Suarez, Bolivia	O 13
583	Porto Velho, Amazonas, Brazil	N 11
584	Portsmouth, Dominica I., Lesser Antilles	N 6
585	Posados, Misiones, Argentina	O 15
586	Potaro Landing, British Guiana	O 8
587	Powdermill Nature Reserve Station near Rector, Westmoreland Co., Penna., USA	K 2
588	Pozuzo, Peru	K 12
589	Prescott, Arizona, USA	D 3
590	Presidio, Texas, USA	F 4
591	Presidio de San Nicolás, Mexico	E 5
592	Princeton, New York, USA	L 2
593	Primavera, Paraguay	O 14
594	Progreso, Yucatán, Mexico	I 5
595	Providencio near Guaymas, Sonora, Mexico	D 4
596	Pucallpa, Peru	K/L 11
597	Puebla, Mexico	G 6
598	Puerto Armuelles, Panama	J 8
599	Puerto Barrios, Guatemala	I 6
600	Puerto Cabello, Venezuela	M 7
601	Puerto Casado, Chaco, Paraguay	O 14
602	Puerto Castilla, Honduras	I 6
603	Puerto Cortes, Honduras	I 6
604	Puerto Isabel, Bolivia	O 13
605	Puerto Leguizamo, Venezuela	L 9/10
606	Puerto Plata, Dominican Republic	L 6
607	Puerto San Lorenzo, Paraguay	O/P 15
608	Puerto Viejo, Chile	L 15
609	Pumpville, Texas, USA	F 4
610	Puna, Ecuador	J/K 10
611	Punta Gorda, British Honduras	I 6
612	Puntarenas, Costa Rica	J 7/8
613	Punta San Blas, Panama	K 8
614	Purulhá, Guatemala	H/I 6/7
615	Puyo, Ecuador	K 10
616	Quaraí, Rio Grande do Sul, Brazil	O 16
617	Quebec, Canada	L 0
618	Quevedo, Ecuador	K 10
619	Quezaltenango, Guatemala	H 7
620	Quincy, Illinois, USA	H 2

621	Quindio Pass, Colombia	K 9
622	Quinigua, Sierra de, Venezuela	M/N 9
623	Quirigua, Guatemala	I 7
624	Quito, Ecuador	K 10
625	Quitobaquito, Organ Pipe National Monument, Arizona, USA	D 3
626	Raleigh, N. Carolina, USA	K 2
627	Ramsey Lake State Park, Fayette Co., Ill., USA	J 2
628	Rancagua, Chile	L 16
629	Rancho Grande near Maracay, Venezuela	M 7
630	Recife, Pernambuco, Brazil	T 11
631	Reno, Nevada, USA	B/C 2
632	Resistencia, Argentina	O 15
633	Restrepo, Colombia	L 9
634	Retalhulen, Guatemala	H 7
635	Reyes, Bolivia	M 12
636	Richmond Hill, Ontario, Canada	K 1
637	Riobamba, Ecuador	K 10
638	Rio Claro, São Paulo, Brazil	Q 14
639	Rio de Janeiro, Rio de Janeiro, Brazil	R 14
640	Rio de Janiero, Federal District Corcovado Forest, Rio de Janeiro, Brazil	R 14
641	Rio Grande, Rio Grande do Sul, Brazil	P 16
642	Rioja, Peru	K 11
643	Rio Negro, Paraná, Brazil	Q 15
644	Rio Pardo, Rio Grande do Sul, Brazil	P 15/16
645	Rio Piedras, Puerto Rico	M 6
646	Rio Prêto, Minas Gerais, Brazil	Q 15
647	Rio Songo, Bolivia	M 13
648	Rio Tapajós, Brazil	O 10
649	Rivera, Uruguay	O 16
650	River Ranch, Florida, USA	J 4
651	Riverside, California, USA	C 3
652	Robore near Chiquitos, Bolivia	N 13
653	Rockaway Park, Long Island, N.Y., USA	L 1
654	Rolândia, Paraná, Brazil	P 14
655	Rosario, Santa Fé, Argentina	N 16
656	Roseau, Dominica I., Lesser Antilles	N 6
657	Rosharon, Texas, USA	G 4
658	Royal Palm Ranger Station, Everglades, Fla., USA	J 5
659	Runaway Bay, Jamaica	K 6
660	Rurrenabaque, Rio Beni, Bolivia	M 12
661	Sabara near Belo Horizonte, Minas Gerais, Brazil	R 13
662	Sagua de Tánamo, Cuba	K 5
663	Salem, Virginia, USA	J/K 2
664	Salesópolis, São Paulo, Brazil	Q/R 14
665	Salina Cruz, Mexico	H 6
666	Salta, Salta, Argentina	M 14/15
567	Saltillo, Coahuila, Mexico	F 4/5
668	Salt Lake City, Utah, USA	D 1/2
669	Salto, São Paulo, Brazil	Q 14
670	Salto Grande, Rio Paranapanema, Brazil	P/Q 14
671	Salvador, Bahía, Brazil	S 12

672	San Andres Tuxtla, Vera Cruz, Mexico	G/H 6
673	San Antonio north of Aliança, Amazonas, Brazil	N 12
674	San Antonio, Coahuila, Mexico	F 5
675	San Antonio, Honduras	I 7
676	San Antonio, Texas, USA	G 4
677	San Benito, Texas, USA	G 4
678	San Bernardino, Paraguay	O 14/15
679	San Carlos, Costa Rica	J 7
680	San Carlos, Santa Cruz, Bolivia	N 13
681	Sanchez, Dominican Republic	M 6
682	San Cristobal, Cuba	J 5
683	San Cristobal de las Casas, Chiapas, Mexico	H 6
684	San Diego, California, USA	C 3
685	San Dimas Canyon, Angeles National Forest, Calif., USA	C 3
686	San Ernesto, Bolivia	M 12/13
687	San Fernando, Trinidad	N 7
688	San Fernando de Apure, Venezuela	M 8
689	Sanford, Florida, USA	J 4
690	San Francisco, Calif., USA	B 2
691	San Francisco de Guadeloupe, Costa Rica	J 7/8
692	San Ignacio, Misiones, Argentina	R 15
693	San Isidro, Costa Rica	J 8
694	San Isidro, Mexico	F 5
695	San José, Catamarca, Argentina	M 15
696	San José, Cordoba, Argentina	N 16
697	San José, Costa Rica	J 7/8
698	San José, Guatemala	H 7
699	San José de Chiquitos, Bolivia	N 13
700	San José del Cabo, Baja California Sur, Mexico	E 5
701	San José Purna, Michoacan, Mexico	F 6
702	San Juan, Dominican Republic	L/M 6
703	San Juancito, Honduras	I 7
704	San Mateo, Costa Rica	J 7/8
705	San Martin, Meta, Colombia	L 9
706	San Martin de los Andes, Patagonia, Argentina	L 18
707	San Miguel de Allende, Guanajuato, Mexico	F 5
708	San Miguel Hidalgo, Mexico	F/G 5
709	San Patricio, Rio Nueves, Texas, USA	G 4
710	San Pedro, Jujuy, Argentina	M/N 14
711	San Pedro Sula, Honduras	I 6
712	San Ramón, Chanchamayo, Peru	K 12
713	San Ramón, Costa Rica	J 6/7
714	San Ramón, Nicaragua	I 7
715	San Salvador, El Salvador	I 7
716	San Salvador, Peru	L 10
717	Santa Barbara, Chihuahua, Mexico	E 4
718	Santa Barbara, Venezuela	M 9
719	Santa Clara, Cuba	J/K 5
720	Santa Cruz, Bolivia	N 13
721	Santa Cruz, Mexico	D 3
722	Santa Cruz, Nayarit, Mexico	E 5
723	Santa Cruz do Sul, Rio Grande do Sul, Brazil	P 15

724	Santa Elena, Ecuador	J 10
725	Santa Elena, Venezuela	M 9
726	Santa Marta, Colombia	L 7
727	Santa Monica (= part of Los Angeles), Calif., USA	C 3
728	Santander, Colombia	K 9
729	Santarem, Pará, Brazil	O/P 10
730	Santa Rosa, Bolivia	M 12
731	Santiago de Chile, Chile	L 16
732	Santiago de Cuba, Cuba	K 5/6
733	Santo Antonio, Mato Grosso, Brazil	O 13
734	Santo Domingo, Carabaya, Peru	M 12
735	Santo Domingo de los Colorados, Ecuador	K 10
736	Santos, São Paulo, Brazil	Q 14
737	São Leopoldo, Rio Grande do Sul, Brazil	P 15/16
738	São Mateu, Expirito Santo, Brazil	R/S 13
739	São Paulo, São Paulo, Brazil	Q 14
740	São Paulo de Olivença, Amazonas, Brazil	M 10
741	Sapucay, Paraguay	O 15
742	Satipo, Peru	L 12
743	Sauk City, Wisconsin, USA	I 1
744	Scott City, Kansas, USA	F 2
745	Serra Alta, Santa Catarina, Brazil	Q 15
746	Serra do Navio, Rio Amapari/Araguari, Brazil	P 9
747	Serro Prieto Mt., Sonora, Mexico	E 4
748	Sete Lagoas, Minas Gerais, Brazil	R 13
749	Sharon, Massachusetts, USA	L 1
750	Shasta, Calif., USA	B 2
751	Shreveport, Louisiana, USA	H 3
752	Sierra Maestra Mts., Cuba	J 5/6
753	Silver Springs, Florida, USA	J 4
754	Simiti, Colombia	L 8
755	Soledad, Santa Clara, Cuba	J 5
756	Solola, Guatemala	H 7
757	Stann Creek, British Honduras	I 6
758	Stanley (= Port Stanley), Falkland I.	O 20
759	St. Anne, Guadeloupe, Lesser Antilles	O 6
760	St. Ann's Bay, Jamaica	K 6
761	St. Augustine, Trinidad	N 7
762	St. George's, Grenada I., Lesser Antilles	N 7
763	St. Jean du Maroni, French Guiana	P 8
764	St. Joseph, Trinidad	N 7
765	St. Laurent du Maroni, French Guiana	P 8
766	St. Petersburg, Florida, USA	J 4
767	St. Pierre, Martinique, Lesser Antilles	N 7
768	St. Thomas, Barbados I., Lesser Antilles	O 7
769	Suffolk, Virginia, USA	K 2
770	Summerdale, Alabama, USA	I 3
771	Suncho Corall, Santiago del Estero, Argentina	N 15
772	Sunflower, Tonto National Forest, Arizona, USA	D 3
773	Tabaconas, N. Peru	K 11
774	Tabasco, Mexico	F 5
775	Taboga, Taboga I.	K 8
776	Tacua, S. Peru	L/M 13

777	Tacuarembo, Uruguay	O 16
778	Tacubaya (= part of Mexico City), Mexico	G 6
779	Tado, Colombia	K 8
780	Tamazunchale, San Luis Potosí, Mexico	G 5
781	Tambillo, Peru	K/L 11
782	Tampico, Mexico	G 5
783	Tancitaro Mt., Michoacan, Mexico	F 6
784	Tandil, Buenos Aires, Argentina	O 17
785	Tapachulo, Mexico	H 7
786	Taperinha, Pará, Brazil	M 10
787	Tarapoto, Peru	K 11
788	Tarija, Bolivia	N 14
789	Tarma, Peru	K 12
790	Taxco de Alarcon, Guerrero, Mexico	G 6
791	Teapa, Tabasco, Mexico	H 6
792	Tefe, Amazonas, Brazil	N 10
793	Tegucigalpa, Honduras	I 7
794	Tehuacan, Mexico	G 6
795	Tehuantepec, Oaxaca, Mexico	G 6
796	Tela, Honduras	I 6
797	Temax, Yucatán, Mexico	I 5
798	Tempe, Arizona, USA	D 3
799	Teocello, Mexico	G 6
800	Teofilo Otoni, Minas Gerais, Brazil	R 13
801	Tepic, Mexico	E/F 5
802	Tepoztlan, Morelos, Mexico	G 6
803	Teresópolis, Rio de Janeiro, Brazil	R 14
804	Tezonapa, Vera Cruz, Mexico	G 6
805	Tijuca (= part of city of Rio de Jan.), Brazil	R 14
806	Timbó, Santa Catarina, Brazil	Q 15
807	Tingo Maria, Peru	K 11
808	Tiquisate, Guatemala	H 7
809	Titusville, Florida, USA	J 4
810	Tlalpujahua, Michoacan, Mexico	F/G 6
811	Tolima Mt., Colombia	K 9
812	Toluca, Mexico	G 6
813	Tonantins, Amazonas, Brazil	M 10
814	Toque-Toque, São Paulo, Brazil	Q 14
815	Toronto, Ontario, Canada	K 1
816	Trancas, Tucuman, Argentina	M 15
817	Trinidad, Cuba	J/K 5
818	Trinidad, Paraguay	O 15
819	Trujillo (= Sto. Domingo), Dominican Republic	L/M 6
820	Trujillo, Peru	K 9
821	Tucavaca, Bolivia	O 13
822	Tucson, Arizona, USA	D 3
823	Tucuman, Tucuman, Argentina	M 15
824	Tuis, Costa Rica	J 8
825	Tujunga Canyon, Angeles National Forest, Calif., USA	C 3
826	Tulsa, Oklahoma, USA	G 2
827	Tumatumari, British Guiana	O 8
828	Tunja, Colombia	L 8

829	Turquino Mt., Sierra Maestra, Cuba	K 5/6
830	Tuxtepec, Oaxaca, Mexico	G 6
831	Uaupes, Rio Negro, Amazonas, Brazil	M 9/10
832	Uberaba, Minas Gerais, Brazil	Q 13/14
833	Urbana, Illinois, USA	L 2
834	Uruapan, Michoacan, Mexico	F 6
835	Urucum, Mato Grosso, Brazil	M 13
836	Valdivia, Chile	L 17
837	Valencia, Venezuela	M 7
838	Valera, Trujillo, Venezuela	L 8
839	Valladolid, Yucatán, Mexico	I 5
840	Valles (=Ciudad Valles), San Luis Potosí, Mexico	G 5
841	Valpareiso, Chile	L 16
842	Venado Tuerto, Santa Fé, Argentina	N 16
843	Venceslau Bráz, Paraná, Brazil	Q 14
844	Ventura, Calif., USA	C 3
845	Vera Cruz, Vera Cruz, Mexico	G 6
846	Verapaz, Guatemala	H 6/7
847	Vernal, Utah, USA	D/E 2
848	Vero Beach, Florida, USA	J 4
849	Vianopolis, Goias, Brazil	Q 13
850	Vicente Lopez (= part of City of Buenos Aires), Argentina	N/O 16
851	Victoria, Texas, USA	G 4
852	Villa Ana, Santa Fé, Argentina	O 15
853	Villa Guillermina, Santa Fé, Argentina	O 15
854	Villa Juarez, Mexico	G 6
855	Villarica, Paraguay	O 15
856	Villa Union, Sinaloa, Mexico	E 5
857	Villavicencio, Colombia	L 9
858	Vitoria, Espirito Santo, Brazil	R 14
859	Volcano Colima, Mexico	F 6
860	Volcano Irazu, Costa Rica	J 7/8
861	Warnes, Bolivia	N 13
862	Washington, D.C., USA	K 2
863	Wilmington, Delaware, USA	K 2
864	Winchendon, Massachusetts, USA	L 1
865	Winnipeg, Manitoba, Canada	G 0
866	Yuma, Arizona, USA	C/D 3
867	Yurimaguas, Peru	K 11
868	Zacualpan, Mexico	G 6
869	Zamora, Ecuador	K 10
870	Zarate, Argentina	N/O 16
871	Zaruma, Ecuador	K 10
872	Zapata, Texas, USA	G 4
873	Zitacuaro, Michoacan, Mexico	F 6

Islands

874	Andros	K 4/5
875	Antigua	N 6
876	Bahia, Honduras	I 6
877	Barbados	O 7

192

878	Barbuda	N 6
879	Barro Colorado, Panama Can. Zone	J/K 8
880	Bonaire	M 7
881	Cayman Brac	K 6
882	Cocos	I 8
883	Coiba, Panama	J 8
884	Dominica	N 9
885	Gorgona, Colombia	K 9
886	Grand Cayman	J 6
887	Grand Turc	L 5
888	Great Abaco	K 4
889	Grenada	N 7
890	Guadeloupe	N 6
891	Hispaniola	L 6
892	La Plata, Ecuador	J 10
893	Little Abaco	K 4
894	Little Cayman	J/K 6
895	Martinique	N 7
896	Mayaguana	L 5
897	Montserrat	N 6
898	New Providence	K 5
899	Perlas, Gulf of Panama	K 8
900	Pinos	J 5
901	Puerto Rico	M 6
902	St. Croix	N 6
903	St. John, Virgin Islands	N 6
904	St. Kitts	N 6
905	St. Thomas	M/N 6
906	St. Vincent	N 7
907	Santa Catarina, Brazil	Q 15
908	Santa Lucia	N 7
909	São Sebastião, Brazil	Q 14
910	Taboga	K 8
911	Tobago	N 7
912	Trinidad	N 7
913	Wakenaam, British Guiana	O 8

Rivers

914	Amapari, Brazil	P 9
915	Araguala, Brazil	P 12
916	Araguari, Brazil	P 9
917	Batalha (near Blumenau), Sta. Catarina, Brazil	Q 15
918	Batalha, São Paulo, Brazil	Q 14
919	Beni, Bolivia	M 12
920	Berbice, British Guiana	O 8/9
921	Blanco, Bolivia	N 12/13
922	Branco, Brazil	N 9
923	Cañete, Peru	K 12
924	Cauca, Colombia	K 8/9
925	Caura, Venezuela	M/N 8
926	Chaparé, Bolivia	M/N 13

927	Chimbo, Ecuador	K 10
928	Chuchungas, Peru	K 11
929	Coari, Amazonas, Brazil	M/N 11
930	Courantym (= Corantijn, Surinam), Brit. Guiana	O 8/9
931	Dagua, Colombia	K 9
932	Demerara, British Guiana	O 8
933	Essequibo, British Guiana	O 9
934	Guaina, Venezuela	M 9
935	Huacamayo, Carabaya, Peru	M 12
936	Huallaga, Peru	K 11
937	Huaura, Peru	K 12
938	Iguaçu, Parana	P 15
939	Inambari, Carabaya, Peru	L/M 12
940	Indian, Florida, USA	J 4
941	Jari, Brazil	P 9
942	Jiparaná (= Machado), Brazil	N 11
943	Jurua, Amazonas, Brazil	N 11
944	Kourou, French Guiana	P 9
945	Kuyuwini, British Guiana	O 9
946	Machado (= Jiparaná), Brazil	N 11
947	Madeira, Brazil	N 11
948	Madre de Dios, Bolivia/Peru	M 12
949	Mamoré, Bolivia	M 12
950	Mana, French Guiana	P 8
951	Mapiri, Bolivia	M 13
952	Mapiri, Bolivia	M 12
953	Maroni, French Guiana	P 9
954	Mazaruni, British Guiana	N/O 8
955	Morona, Peru	K 10
956	Napo, Ecuador/Peru	L 10
957	Nazas, Mexico	F 4
958	Negro, Bolivia	N 12/13
959	Negro, Brazil	N 10
960	Negro, Paraná, Brazil	P/Q 15
961	New River, British Guiana	O 9
962	Orinoco, Venezuela	M 8
963	Oronoque, British Guiana	O 9
964	Oyapok, French Guiana	P 9
965	Pacaya, Peru	K/L 11
966	Pachitea, Peru	L 11
967	Palcazu, Peru	K 12
968	Paulo, Santa Catarina, Brazil	Q 15
969	Paranapanema, Brazil	P 14
970	Paru, N. Brazil	P 9/10
971	Pastaza, E. Ecuador/Peru	K 10
972	Pativilca, Peru	K 12
973	Perene, Peru	K 12
974	Pichis, Peru	K/L 12
975	Pisco, Peru	K 12
976	Potaro, British Guiana	O 8
977	Purus, Brazil	M 11
978	Putumayo, Peru/Colombia	L 10
979	Rapulco, Bolivia	M 12

980	Salado, Argentina	N/O 17
981	San Martin, Bolivia	N 12
982	Solimoés (Amazon)	M 10
983	Suapure, Venezuela	M 8
984	Suriname, Surinam	O 9
985	Tabaconas, N. Peru	K 11
986	Tapajós, Brazil	O 11
987	Tapiche, Peru	L 11
988	Tefe, Amazonas, Brazil	M 10
989	Tocantins, N. Brazil	Q 10/11
990	Tucavaca, Bolivia	O 13
991	Uaupes, Brazil	M 9/10
992	Ucayali, Peru	K/L 11
993	Urubamba, Peru	L 12
994	Velhas, Minas Gerais, Brazil	R 13
995	Verde, Bolivia/Brazil	N 12
996	Vermelho, Goias, Brazil	P 13
997	Vermelho, Goias, Brazil	Q 13
998	Wanks, Nicaragua/Honduras	J 7
999	Xingu, Brazil	P 11
1000	Yacuma, Bolivia	M 12
1001	Yaguas, Peru	L 10
1002	Yavari (= Javari), Amazonas, Brazil	L 10/11

Addenda

1003	Iquitos, Peru	L 10
1004	Chanchamayo, Prov. of, Peru (cf. no. 392)	K 12
1005	Cruzeiro, Rio de Janeiro, Brazil	Q/R 14
1006	Mérida, Yucatán, Mexico	I 5